深入浅出 React.js
原理与实战

冯 昕 编著

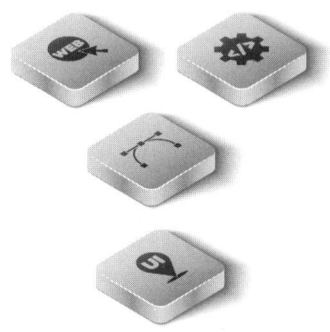

化学工业出版社

·北京·

内容简介

本书以丰富的实例对 React 的底层原理与应用进行了细致的讲解，具体内容包括 React 的快速入门、React 组件的状态、通信与生命周期、React 路由、React Hooks、状态管理、基于 Nest 或 Midway 的全栈化方案，以及两个 React 综合实战项目——搭建 React 组件库和 UI 自动化测试平台。

本书结构清晰，内容由浅入深，适合任何对 React 感兴趣，并计划深入学习 React 的读者。

图书在版编目（CIP）数据

深入浅出 React.js：原理与实战 / 冯昕编著.
北京：化学工业出版社，2025. 7. -- ISBN 978-7-122-48159-7

Ⅰ．TP393.092.2；TP312.8

中国国家版本馆 CIP 数据核字第 20256BR369 号

责任编辑：张　赛　　　　　　　文字编辑：张　宇
责任校对：杜杏然　　　　　　　装帧设计：王晓宇

出版发行：化学工业出版社
　　　　　（北京市东城区青年湖南街 13 号　邮政编码 100011）
印　　装：北京云浩印刷有限责任公司
710mm×1000mm　1/16　印张 18　字数 333 千字
2025 年 7 月北京第 1 版第 1 次印刷

购书咨询：010-64518888　　　　售后服务：010-64518899
网　　址：http://www.cip.com.cn
凡购买本书，如有缺损质量问题，本社销售中心负责调换。

定　　价：89.00 元　　　　　　　版权所有　违者必究

前言
Preface

React 是一个用于构建用户界面的开源 JavaScript 库。其提供的组件化开发模式、虚拟 DOM 等为交互式用户界面的开发提供了自由、快捷、高性能的开发体验，因而在 Web 开发中得到了广泛的认可，并成为众多企业级应用的首选。

对比众多的 React 教程资源，本书并不局限于 React 的简单使用，还关注 React 相关技术的实现原理，通过深挖源码的设计逻辑，并结合实践经验与丰富示例，帮助读者更直观地理解并实现应用。通过本书，读者将系统性地了解 React 的方方面面，从基础知识到进阶技巧，从路由管理到状态管理，再到全栈开发，帮助读者全面理解并掌握 React。

以下是本书各章的主要内容。

第 1 章：主要介绍前端框架的一些基础知识，以及 React 的优势及特点等。读者将通过简单的实例快速了解 React 的使用，以及 ReactDOM 相关知识。

第 2 章：围绕 React 组件相关知识展开，包括组件状态与通信、生命周期，以及 React 的事件机制等内容。读者将了解到类组件和函数组件的定义、使用方法和最佳实践。

第 3 章：介绍如何在 React 应用中使用路由，以及路由的实现原理、路由守卫等内容。

第 4 章：介绍 React 16 版本之后的核心概念——Hook，读者将会学习到 React 的基本内置 Hook、主流开源库的 Hook，以及如何自定义地实现一些常用 Hook。

第 5 章：介绍 React 项目中主流的状态管理库，尤其是基于 Redux 的 react-redux 的应用。

第 6 章：主要讲解如何基于 Nest.js 和 Midway.js 成为一个全栈开发者，包含一些后端的知识以及前后端联调的知识点。

第 7 章：基于 React 的两个综合实战项目——React 组件库和 UI 自动化测试平台。

希望通过本书的学习，读者不仅能掌握 React 的核心思想与技术细节，还能够培养独立解决问题的能力，为将来的职业生涯打下坚实的基础。

前端开发领域发展迅速，由于编者的水平与认识存在局限性，书中难免存在疏漏，敬请读者批评指正。

编著者

目录
Contents

第 1 章
React 快速入门   001

1.1	传统前端开发到现代前端架构	001
1.2	使用 React 的理由	002
	1.2.1 虚拟 DOM	004
	1.2.2 diff 算法	007
	1.2.3 key	010
1.3	ReactDOM	015
	1.3.1 react-dom 与 react 的关系	017
	1.3.2 组件中的状态响应	019
	1.3.3 React 基本引入方式	020
	1.3.4 开箱即用的 React 引入	022
	1.3.5 Vite 快速初始化	023
思考题		025

第 2 章
组件与视图渲染   027

2.1	React 组件初探	027
	2.1.1 类组件	028
	2.1.2 函数组件	030
2.2	组件状态与通信	030
	2.2.1 state	031
	2.2.2 props	036
	2.2.3 props 导致的更新	040

	2.2.4 父传子通信	045
	2.2.5 子传父通信	052
	2.2.6 兄弟组件通信	053
	2.2.7 跨组件分层通信	057
2.3	组件生命周期	061
	2.3.1 类组件生命周期	061
	2.3.2 函数组件生命周期	066
2.4	遍历渲染	067
	2.4.1 遍历渲染对象	067
	2.4.2 遍历渲染数组	069
2.5	React 事件机制	071
思考题		073

第 3 章
React Router   074

3.1	配置路由	077
3.2	React Router 实现原理	079
3.3	React Router V6 详解	085
3.4	路由守卫	091
3.5	哈希路由和历史路由	095
思考题		097

第 4 章
React Hooks 深入浅出   098

4.1	useState	101
4.2	useState 是同步的还是异步的？	106
4.3	useEffect	110
4.4	useLayoutEffect	112
4.5	useEffect 和 useLayoutEffect 的区别	113
4.6	useRef	120

4.7	useMemo 和 useCallback	124
4.8	useContext	127
4.9	useReducer	130
4.10	自定义 Hook	135
	4.10.1　功能型 Hook	137
	4.10.2　业务型 Hook	140
	4.10.3　实现一个完整版 useRequest	143
4.11	ahooks 入门	158
思考题		164

第 5 章
React 状态管理方案　　　　　　　　　165

5.1	主流状态管理方案	165
5.2	Redux	170
5.3	react-redux	174
5.4	实现一个简易版 Redux	178
思考题		184

第 6 章
全栈化与 Serverless　　　　　　　　　185

6.1	Nest.js 快速入门	186
6.2	数据库连接和初始化	208
6.3	快速构建 CRUD	213
6.4	基于 Nest.js 的 RBAC 权限控制系统实现	216
6.5	JWT 登录及伪造请求解决方案	224
6.6	跨端扫码登录	237
6.7	Midway.js 入门	242
6.8	Midway.js 实现注册、登录、鉴权	252
思考题		262

第 7 章
企业级 React 项目实战   263
7.1 搭建 React 组件库   263
7.2 搭建 UI 自动化测试平台   269
思考题   279

结语   280

第 1 章
React 快速入门

React 是一款高效灵活的现代化前端框架，它采用了组件化的开发思想，将复杂的用户界面分解为多个独立、可复用的组件，使得代码结构更加清晰、易于维护和扩展。也正是因此，对于需要构建大型、复杂的页面应用的开发人员来说，React 即为最佳选择。

1.1 传统前端开发到现代前端架构

在互联网发展的早期，网页的交互性和动态更新主要依赖于原生 JavaScript 和 JQuery，开发者们在这两个技术的支持下实现了基本的用户交互、表单验证和 DOM 操作。然而，随着技术的不断发展，用户对于网页应用的期望和要求也与日俱增，传统的开发方式逐渐体现出了其局限性。

例如早期应用较多的多页应用（MPA）模型，其在每次请求时都会重新加载整个页面，用户体验较差。在此背景下，单页应用（SPA）问世，面对同样的请求场景，其在请求响应后只更新页面的一小部分，从而极大地提升用户体验。而在 SPA 的实现过程中，不同的架构模式也应运而生。

- MVC（Model-View-Controller）：将应用分为模型（Model）、视图（View）和控制器（Controller）。在 MVC 中，控制器负责更新视图，也就是请求之后的页面更新；模型则用于存储应用的状态。尽管 MVC 提供了良好的分离关注点的能力，但随着应用需求的复杂度增加，管理模型与视图之间的交互就会变得更复杂。
- MVVM（Model-View-ViewModel）：MVVM 进一步简化了这一过程，其在视图与模型之间增加了一层绑定，即通过 ViewModel 将模型与视图绑定，提供用户接口逻辑，从而减轻控制器的负担。但 MVVM 也并非完美，由于多了一层架构，在面临数据流的复杂情况时，其调试也会变得更加困难。在实际开发过程中，笔者发现 MVVM 相较于 MVC，代码量并不降反增。但另一方面，因其多了一层数据绑定的关系逻辑，逻辑看上去会清晰很多，对于整体系统的可读性还是有不少优势的。

随着 SPA 和现©代前端架构的发展，为了解决 MVC 和 MVVM 在复杂应用中的问题，React 采用了组件化的设计理念，将用户界面拆分为独立、可复用的组件。每一个组件都有自己的独立状态和逻辑，使得管理复杂的用户界面变得更简单。

React 的核心机制主要有两个：

- 虚拟 DOM：通过在内存中维护一个与真实 DOM 相对应的虚拟 DOM 树，React 在组件状态发生变化时，首先在虚拟 DOM 上操作，通过差异算法找出变化并且一次性更新到真实 DOM 上，减少直接操作 DOM 的成本，提升性能；
- 单向数据流：React 遵循单向数据流的原则，数据从父组件传递到子组件，这使得数据管理更加清晰，状态比较容易追踪。

与传统的 JavaScript、JQuery 开发相比，React 的性能优势主要体现在：

- 高效的 DOM 更新：通过虚拟 DOM 和 diff 算法，React 能大幅度降低 DOM 更新次数；
- 可复用的组件：在复杂应用中，可以通过组件化的技术将重复的逻辑和模块抽离出来，这样做不仅提升了代码的复用性，还减少了代码冗余和维护成本；
- 强大的状态管理：通过 useState 和 useEffect 等钩子函数（Hook），React 使得状态管理更加简单、逻辑更为清晰，在大型应用中也能更有效地处理组件之间的逻辑。

1.2 使用 React 的理由

在此之前，你可能接触了比较久的 JQuery 或者原生的编程范式，但如果你对于浏览器渲染机制有一定认识，那你一定会知道频繁地操作 DOM 元素对于网页性能而言损耗是非常大的，就像是这样：

```html
<body>
  <h1 class="tag">初级前端工程师</h1>
  <span>我的技能是 JQuery</span>
  <button onclick="click">学习</button>
  <script>
    function click() {
      document.querySelector('.tag').innerText='高级前端工程师';
      document.querySelector('span').innerText='我的技能是 React';
    }
  </script>
</body>
```

这段代码实现了一个非常简单的小功能：点击按钮改变两个元素的文案。其对两个 DOM 元素进行了直接操作，由于两个元素同步单线程变化，页面进行了两次重绘。如果以类似方案实现更复杂的业务需求和交互，比如点击一次按钮更新一万条数据，那么后果可能不堪设想，页面性能会变得非常差。

在 React 中，由于其是一个视图层框架，将数据层和视图层完全分离，让开发者只需要专注于数据的改变，当数据（JS）发生变化时，网页（HTML）也会自动发生变化。

同样的功能放在 React 中是这样的：

```
import { useState } from 'react';

const ReactComponent = () => {
  const [job, setJob] = useState('初级前端工程师');
  const [skill, setSkill] = useState('JQuery');
  const click = () => {
    setJob('高级前端工程师');
    setSkill('React');
  }
  return (
  <div>
    <h1 classNmae="tag">{job}</h1>
    <span>我的技能是{skill}</span>
    <button onClick={click}>学习</button>
  </div>
  )
}

export default ReactComponent;
```

这段代码是 React 的基础写法，在这段代码中我们做了这些事情：
- 定义了一个名为 ReactComponent 的 React 函数（组件）；
- 函数中声明了两个数据（React 状态）和改变其对应的方法（更新）；
- 将在函数中声明的 React 数据组装在 HTML 元素中作为函数返回值；
- 导出这个 React 函数。

这就是当下 React 的常规写法。这段代码块中点击事件触发后并没有看到直接操作 DOM 元素的 API，而是调用了 useState 解构出来的两个函数，这样我们的页面就刷新了，而在这过程中虽然调用了两次方法，但其在 React 内部实现中是异步更新的，因而只进行了一次页面重绘。

至此，你应该已经了解这两种实现最直观的区别了。接下来介绍一下使用 React

第 1 章 React 快速入门　003

的优点。

- 高性能。React.js 的写法其实和原生 JS 并没有很大的差异，它基于 DOM，这是它非常快的原因，并且 DOM 的变化被最小化，其针对性能和速度进行了优化，减轻了 CPU 的负担。
- 生态齐全。React 诞生早且易于扩展，其庞大且完善的生态系统可以帮助开发者解决各种问题，提升开发效率，因此使用 React 可以获得相比于其他框架更完备的技术选型。
- 社区支持。在国内，目前使用 Vue 的人员可能也很多，但是面向世界，React 的使用比例是非常庞大的。如果你在开发中遇到问题，你可以在线上找到绝大多数问题的解决方案。
- 组件复用。你一定听说过函数封装、代码抽离、减少代码冗余。对于通用逻辑，你可以封装成组件来开发，原理是一样的，这样将极大地减少对于 HTML 层面的代码冗余，并且更易于项目维护。
- 虚拟 DOM。React 之所以可以做到批量更新，背后的原理其实是高效的虚拟 DOM。这是一种将 DOM 存储在内存中的概念，其将所有更改存储起来一次性进行页面更新，这正是 React 高性能的核心。
- 易于理解。React 是一个非常容易学习和理解的 JavaScript 库，你甚至可以在短时间内快速掌握并开发出一个 TODO List 页面。

1.2.1 虚拟 DOM

在传统的 Web 应用中，我们往往会把数据的变化实时地更新到用户界面中，因而每一次改动都引发页面的重渲染，此前提到的多次更新数据层的场景便是如此。而 React 通过异步处理只进行一次页面重绘，其中的奥秘就在于虚拟 DOM。虚拟 DOM 的目的是将过程中所有的改动都积攒在一起，然后一次计算出所有的变化，统一更新一次 DOM。

例如有这样一个 HTML 模板：

```
<div class="title">
  <span>Hello React</span>
  <ul>
    <li>苹果</li>
    <li>橘子</li>
  </ul>
</div>
```

其在 React 中会被转换为这样的 JS 对象：

```
const vitrualDom = {
  type: 'div',
  props: { class: 'title' },
  children: [
    {
      type: 'span',
      children: 'Hello React',
    },
    {
      type: 'ul',
      children: [
        { type: 'li', children: '苹果' },
        { type: 'li', children: '橘子' },
      ],
    },
  ],
};
```

当需要创建或更新视图层时，React 会先通过模板创建出虚拟 DOM 对象，最后再将该对象转换为真实 DOM（此时视图更新）。当然，在实际的 React 虚拟 DOM 中不止这些基本类型，React 会通过 React.createElement 的方式将整个模板转换成虚拟 DOM 对象，例如上述案例最终生成的虚拟 DOM 对象如图 1.1 所示。

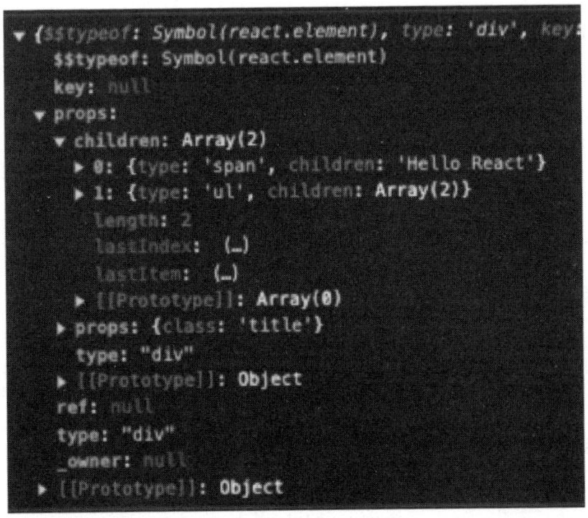

图 1.1

图 1.1 体现了虚拟 DOM 的一些基本的数据结构，作用如下：

● typeof: 元素类型，可以是原生元素（span/ul/div）、文本或 React 组件（ReactComponent）；

- key：每个元素的唯一标识，用于 Diff 算法比较页面元素更新前后差异；
- ref：访问原生 DOM 节点，在 React 中用于特殊情况手动操作 DOM；
- props：元素传递参数，如类名、innerHTML，也可以是一个新的元素（嵌套元素），或者是 React 组件；
- owner：指示当前正在构建的节点属于哪个 React 组件；
- self：指定位于哪个组件实例（非生产环境）；
- _source：指示元素代码来自于哪个文件以及代码行数，相当于 sourcemap（非生产环境）。

可见，虚拟 DOM 本身其实并不稀奇，只是一个 JS 对象里面又嵌套了许多 JS 对象，就像是一个树节点。在 React 中，组件模板只能有一个根元素，这样就是一个唯一父节点向下向内扩展的数据结构。如果有多个根节点，情况则会变得非常复杂，不利于框架本身的维护。

如果页面结构复杂，则再通过 React.createElement 的方式去转换虚拟 DOM 对象，对于开发者而言会变得难以接受。因此，React 官方推出了 JSX 语法（JS 的扩展），令 JavaScript 语言获得了模板化，用户只需要像下方代码一样去开发，React 便会通过 JSX Babel 插件将 html 元素转换为 React.createElement。

```
const Index = () => {
  return (
    <div>
      你好
      <p>欢迎</p>
      <Children>这是子组件</Children>
    </div>
  );
};
```

上面的 React JSX 模板会被 Babel 解析成 React.createElement 版本的形式：

```
const Index = () => {
  return React.createElement(
    'div',
    null,
    '你好',
    React.createElement('p', null, '欢迎'),
    React.createElement('Children', null, '这是子组件'),
  );
};
```

可见，以 JSX 来代替 React.createElement 的写法将大大降低开发者的负担，使我们可以专注于业务本身。

顺便一提，为什么 React 中的组件首字母必须要大写呢？这是由于通过 Babel 对 JSX 进行转换时，需要一个条件（即首字母大小写）来判断该节点是 React 组件还是原生标签。

1.2.2 diff 算法

对于 React 多次更新只进行一次重绘，你可能会思考这个问题：React 是如何进行更新前后的对比，最后进行唯一一次更新的呢？

对此，React 会维护两个虚拟 DOM 树。那么，如何来对比、判断，做出最优的解法呢？这就用到了 diff 算法。

在 React 中，最值得称赞的就是虚拟 DOM 和 diff 算法的结合，React 在引入 diff 算法的基础上还对其进行了质的优化。例如在计算一棵树到另一棵树有哪些节点发生变化时，传统 diff 算法会通过循环+递归的形式去依次对比，并且该过程不可逆，时间复杂度达到了 $O(n^3)$。简单理解，如果有一千个节点，就需要计算十亿次！而 React 的 diff 算法的时间复杂度只有 $O(n)$，这是非常巨大的优化。该算法基于以下核心假设：

- Web UI 中 DOM 节点跨层级的移动操作特别少，可以忽略不计；
- 拥有相同类的两个组件将会生成相似的树形结构，拥有不同类的两个组件将会生成不同的树形结构；
- 对于同一层级的一组子节点，它们可以通过唯一 ID 进行区分。

React 的 diff 算法主要通过三种策略实现：tree diff、component diff、element diff。

（1）tree diff（同级比较）

因为 DOM 节点跨层级的移动少到可以忽略不计，React 便通过 updateDepth 对虚拟 DOM 进行层级控制，在同一层如果发现某个子节点不在了，会完全删除并中断向下的比较，这样只遍历一次即可。

如图 1.2 所示，在 diff 比较时会一层一层地比较：第一层根节点一致，到第二层的时候会发现 R 节点还在并且属性和节点类型都一致，但左侧的树的 L 节点消失了，此时 React 会直接删除 L 节点，重新创建。

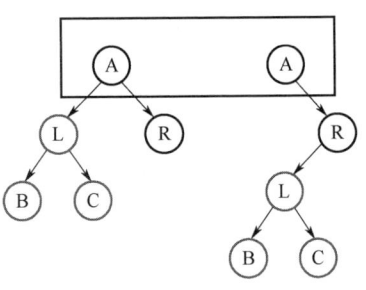

图 1.2

（2）component diff（组件比较）

对于 React 目前策略一共有两种：对于同类型组件，进行常规的虚拟 DOM 判断即可（但是会有特殊情况，例如当组件 A 变化到组件 B 的时候，虚拟 DOM 没有发生任何变化，所以用户可以通过 shouldComponentUpdate 来决定是否需要更新和计算）；对于不同类型组件，例如组件名不同、组件类型不同（class → function），React 会统一认定为 dirty component（脏组件），会直接删除该组件进行重新创建，不做 diff 比较。

如图 1.3 所示，当比较 D → G 时，虽然两个组件的结构非常相似，但是 React 判定 D 为 dirty component，就会直接删除 D，重新构建 G。两个组件不同，结构类似的情况也是非常少的。

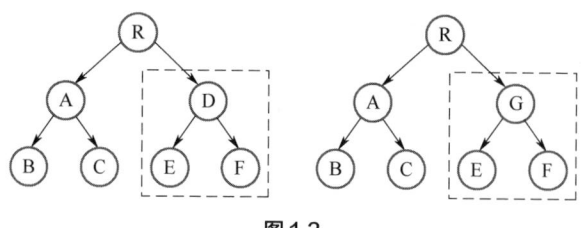

图 1.3

（3）element diff（节点对比）

对于同一层级的节点进行唯一 key 的比较，当所有的节点在同一层级时包含三种操作：插入（INSERT_MARKUP）、移动（MOVE_EXISTING）、删除（REMOVE_NODE）。

- INSERT_MARKUP：旧的层级没有，但新层级中有的节点，直接插入新的层级中。如 C 不在 A、B 中，直接插入。
- MOVE_EXISTING：旧的层级中有，且在新层级中也有的节点，属性相同、可以复用的情况下，采取移动的方式复用以前的 DOM 节点。如旧层级为 A、B、C、D，新层级为 A、D、C、B，节点相同，只是位置发生了变化，这种情况只需要复用和移动即可。
- REMOVE_NODE：旧层级中有，但新层级中没有的节点，直接删除。或者新旧节点属性发生变化，无法复用，也会执行删除。如旧层级为 A、B、C、D，新层级为 A、B、C，则直接删除 D 节点。

以下以两个实际场景进行举例。

场景一：如图 1.4 所示，相同节点，移动复用。

① React 开始遍历新层级，遇到 B 节点，判断其在旧层级中是否出现（发现 B 节点），判断是否要移动 B；

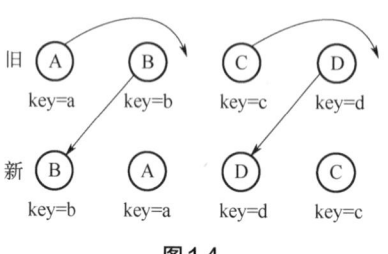

图 1.4

② 判断一个节点是否移动的条件为 index < lastIndex，index 为旧层级的下标，lastIndex 为新层级中的下标或者移动后的下标，取 Math.max(index, lastIndex)，对于 B，index=1、lastIndex=0，因此不满足移动条件；

③ 此时到了 A 的比较，index=0，lastIndex=1，满足 index < lastIndex 的条件，移动 A 节点，lastIndex 为 1；

④ 相同的方式到了 D 节点，index=3，lastIndex=1，不满足 index < lastIndexDe 条件，不移动，并且 lastIndex 更新到 3；

⑤ 最后到了 C 节点，index=2，lastIndex=3，满足条件，移动 C，lastIndex 还是 3，此时该层级 Element diff 结束。

场景二：如图 1.5 所示，有新节点加入，删除节点。

① B 与场景一一样，不移动，lastIndex 为 1；

② 发现没有 E 节点，直接创建，此时 lastIndex 还是 1；

③ 到 C 节点，index=2，lastIndex=1，不满足 index < lastIndex 的条件，不移动 C，lastIndex 更新到 2；

④ 到 A 节点，index=0，lastIndex=2，index < lastIndex，移动 A 节点；

⑤ 不存在 D 节点，删除 D 节点。

再来看一种特殊情况：如图 1.6 所示，走到 D 节点，index=3，lastIndex=0，D 节点被移动到了第一位，导致 A、B、C 节点都替换了位置，影响了所有的 DOM 性能，因此开发中要减少尾部节点移动到首部的操作。

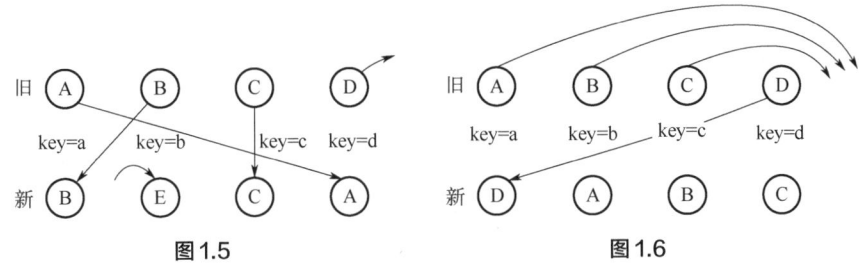

图 1.5　　　　　　　　　　　　图 1.6

还有一个热门话题：为什么遍历时不要用 index 作为节点的 key？对此，假设有一个长度为 100 的数组，如果在 index 为 10 的时候破坏了后续数组的排列，比如插入或删除了一个节点，长度变成了 99 或 101，这样后续的节点 index 全部发生了变化（图 1.7）。

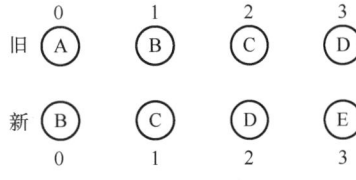

图 1.7

该过程的代码如下所示。

```
const Index = () => {
  const [list, setList] = useState(new Array(100).map((_, index) =>
index));
  const change = () => {
    setList((old) => {
      old.splice(10, 1, 'new');
      return [...old];
    });
  };
  return (
    <>
      {list.map((_, index) => {
        return <div key={index}>{index}</div>;
      })}
      <button onClick={change}>改变</button>
    </>
  );
};
```

实际上，节点并没有发生变化，因此在某次遍历或整个数组的数据变化（如分页）中，使用 key 是可以接受的。然而，在其他情况下，应该使用唯一标识作为 key。否则如果数组长度较长，这样的做法可能导致显著的性能损耗。

1.2.3 key

在 1.2.2 diff 算法中其实讲到了一部分 key 的作用，而 key 最大的作用就是标识一个元素的唯一性。如果在 diff 时 key 发生了变化，就会直接重新创建新的元素。然而正常情况下并不是每个元素都需要 key 的标识，只有在循环渲染的时候，才需要定义 key 来标识元素。例如你写了这段代码：

```
const data = [
  { id: 0, name: 'list1' },
  { id: 1, name: 'list2' },
  { id: 2, name: 'list3' },
  { id: 3, name: 'list4' },
];
const ListItem = (props) => {
  return <li>{props.name}</li>;
};
const List = () => {
  return (
    <ul>
```

```
    {data.map((item) => (
      <ListItem name={item.name} />
    ))}
  </ul>
  );
};
```

你会发现在控制台出现了 React 所提示的警告信息，如图 1.8 所示。

1. Each child in a list should have a unique "key" prop.

图 1.8

该警告信息提示：列表的每一个子元素都应该有一个唯一的 key 属性。在循环渲染的 ListItem 组件中加入 key 属性即可消除这个警告：

```
const data = [
  { id: 0, name: 'list1' },
  { id: 1, name: 'list2' },
  { id: 2, name: 'list3' },
  { id: 3, name: 'list4' },
];
const ListItem = (props) => {
  return <li>{props.name}</li>;
};
const List = () => {
  return (
    <ul>
      {data.map(({name, id}) => (
        <ListItem name={name} key={id} />
      ))}
    </ul>
  );
};
```

在 React 的 diff 算法中，元素 key 属性的作用是判断该元素是新创建的还是被移动的元素，从而减少不必要的 diff。因此在循环场景下，每个元素需要一个可确定唯一的标识。以上述代码为例，将其改造成一个可迭代、向 data 尾部插入数据的代码块：

```
const data = [
  { id: 0, name: 'list1' },
  { id: 1, name: 'list2' },
```

```
  { id: 2, name: 'list3' },
  { id: 3, name: 'list4' },
];
const ListItem = (props) => {
  return <li>{props.name}</li>;
};
const List = () => {
  const [listData, setListData] = React.useState(data);
  const addData = () => {
    setListData((old) => {
      old.push({
        id: old.length,
        name: `list${old.length + 1}`,
      });
      return [...old];
    });
  };
  return (
    <>
      <ul>
        {listData.map(({ name, id }) => (
          <ListItem name={name} key={id} />
        ))}
      </ul>
      <button onClick={addData}>尾部 push 一条数据</button>
    </>
  );
};
```

当每次点击按钮时,新旧节点与 key 的映射图如图 1.9 所示。

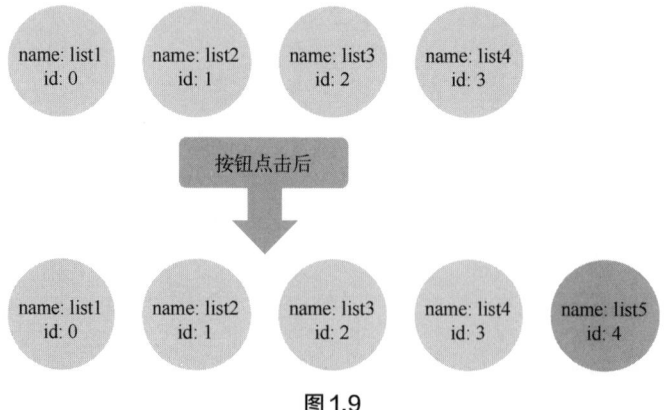

图 1.9

可以看到，在数组尾部插入新数据的情况，新增元素 key 在递增，原有元素节点的 key 在点击按钮重渲染后并没有发生变化（当然这也是预期的），因此四个原有节点在第一次点击按钮后复用了，只进行了一次新元素的添加，也是性能较理想的一次 diff。

但是如果换一种方式呢？比如向数组中间插入一个元素节点呢？将代码做以下调整：

```
const data = [
  { id: 0, name: 'list1' },
  { id: 1, name: 'list2' },
  { id: 2, name: 'list3' },
  { id: 3, name: 'list4' },
];
const ListItem = (props) => {
  return <li>{props.name}</li>;
};
const List = () => {
  const [listData, setListData] = React.useState(data);
  const addData = () => {
    setListData((old) => {
      old.splice(0, 2, {
        id: old.length,
        name: `list${old.length + 1}`,
      })
      return [...old];
    });
  };
  return (
    <>
      <ul>
        {listData.map(({ name, id }) => (
          <ListItem name={name} key={id} />
        ))}
      </ul>
      <button onClick={addData}>在第二位插入一条数据</button>
    </>
  );
};
```

当点击按钮在中间插入一条数据后，映射如图 1.10 所示。

看起来似乎也没什么问题，在原数组的第三位插入了一个新节点，key=4，其他四个节点的 key 没有发生改变，因此得到了复用。

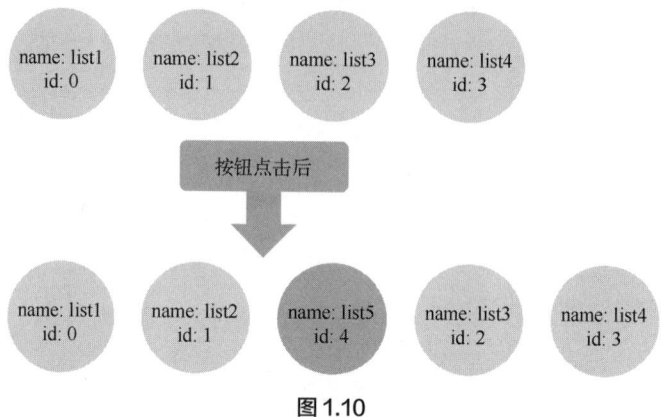

图 1.10

那么如果采用数组下标 index 来作为 key 呢？结果变成了这样：

```
const data = [
  { id: 0, name: 'list1' },
  { id: 1, name: 'list2' },
  { id: 2, name: 'list3' },
  { id: 3, name: 'list4' },
];
const ListItem = (props) => {
  return <li>{props.name}</li>;
};
const List = () => {
  const [listData, setListData] = React.useState(data);
  const addData = () => {
    setListData((old) => {
      old.splice(0, 2, {
        id: old.length,
        name: `list${old.length + 1}`,
      })
      return [...old];
    });
  };
  return (
    <>
      <ul>
        {listData.map(({ name, id }, index) => (
          <ListItem name={name} key={index} />
        ))}
```

```
      </ul>
      <button onClick={addData}>在第三位插入一条数据</button>
    </>
  );
};
```

对应的节点 diff 映射图如图 1.11 所示。

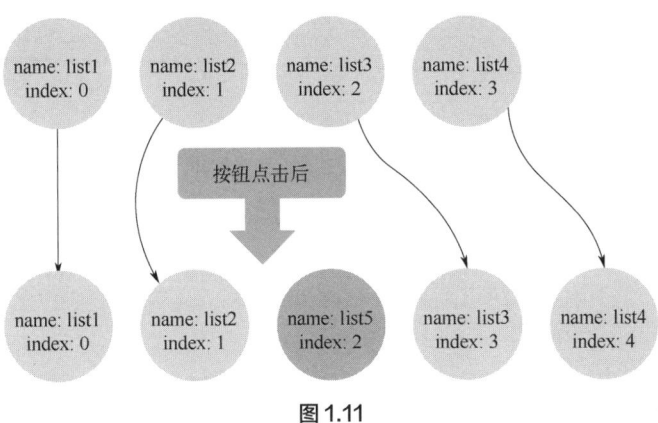

图 1.11

你是否发现了什么问题？为什么 list3、list4 节点明明没有变化，只是被往后推了一位，但却重新渲染新建节点且并没有复用？这是由于使用 index 作为 key，当数组顺序被破坏后，重渲染时 index 也发生了变化，因此重新插入的 list5 开始往后的所有节点全部都无法复用，只能重新创建了,这样极大地降低了 diff 的效率。

如果你理解了这个 list 案例中的奥秘，相信在未来 React 应用开发的路上，key 是不会成为你提升应用性能的瓶颈的。至此，总结出 key 的注意事项：

- key 应该是唯一的；
- key 不要使用随机值，随机数每一次 render 都不一样，节点每次都无法复用；
- 避免使用 index 作为 key。

1.3 ReactDOM

当构建一个 React 应用程序时，入口文件（App.js）看起来像是这样的：

```
import React from 'react';
import { createRoot } from 'react-dom/client';
const App = () => {
return (
<div>
```

```
<h1>Hello world!!</h1>
</div>
);
};
const container = document.getElementById('root');
const root = createRoot(container);
root.render(<App />);
```

上述代码分别引入了 react 和 react-dom，主要是因为 React 不仅能用在 Web 页面，还能用在服务端 SSR、移动端、桌面端，而 ReactDOM 只是运行在 Web 页面（H5）的页面交互包。

React 在 v0.14 之前是没有 react-dom 的，所有功能都包含在 react 中。React 能兼容更多的终端，是从 v0.14（2015-10）开始的，react 库被拆分成了 react 和 react-dom。React 团队的视野首先扩展到了 react-native，react 只包含了 Web 和 Native 通用的核心部分，因此你可以理解 react-native 是在 Native 环境中的 react-dom。可见，若 react 被命名为 react-core 可能会更直观些。

举个例子，下面是一个 Web 端的 React 项目入口根文件（你很难在项目非入口文件中看到 react-dom 的引用）：

```
import React from 'react';
import { createRoot } from 'react-dom/client';
const App = () => {
return (
<div>
<h1>Hello world!!</h1>
</div>
);
};
const container = document.getElementById('root');
const root = createRoot(container);
root.render(<App />);
```

再看一段移动端的 react-native 代码：

```
import React from 'react';
import { Text, View } from 'react-native';
const App = () => {
  <View>
    <Text>Hello world!!</Text>
  </View>
}
```

可以发现，两者的差别就是实现 DOM 交互的引入库发生了变化，这也是为什么有过 React 基础的开发者可以快速上手 react-native 的原因，因为他们的编程范式是相通的——组件化、Hook、MVC 架构思想等。

1.3.1 react-dom 与 react 的关系

简单总结一下：react 负责描述特性、提供 React API，例如类组件、函数组件的定义，基于函数式组件的内置 Hook、refs，基于类组件的内置生命周期、钩子（Hook）函数等。而 react-dom 负责基于 react 所提供的特性（范式）来实现整个特性和规则，并渲染到页面中，因此 react-dom、react-native 也称为渲染器，负责在不同的运行环境上实现 react 通配的特性，达到原生真实效果，例如在浏览器中渲染 DOM 树、响应各种交互事件等。

此时再回到这段代码块：

```
import React from 'react';
import ReactDOM from 'react-dom';
const App = () => {
  <div>
    <h1>Hello world!!</h1>
  </div>
}
ReactDOM.render(<App />, document.getElementById('root'));
```

入口显式调用了 ReactDOM.render，入参为一个组件和一个对象，从结果层面来看，这次函数调用最终将 DOM 树挂载到了节点上（root），那么这个根组件实例是什么呢？其实就是 1.2.2 节所讲到的虚拟 DOM，一个 vDOM 对象，由 React.createElement 函数创建，而 vDOM 也是 react 和 react-dom 之间最直观的联系，具体的数据结构如果有遗忘，可以翻阅重顾。

再看一个以 App 组件为根节点且包含子节点的例子：

```
import React from 'react';
import './styles.css';
const ComponentA = () => {
  return <h1>Component A</h1>;
};
const ComponentB = () => {
  return <h2>Component B</h2>;
};
const App = () => {
  return (
    <div>
```

```
      <ComponentA />
      <ComponentB className={'componentB-style'} />
    </div>
  );
};
export default App;
```

这段代码中定义了一个 App 组件，并且包含了两个子组件，分别是 ComponentA 和 ComponentB，最终向外暴露 App 组件，进行 ReactDOM.render 的渲染。我们把 App 组件的数据结构打印出来看一下：

```
console.log(<App />.type());
// 输出：
{
  type: 'div',
  key: null,
  props: {},
  className: 'App',
  children: [
    {
      type: function ComponentA(){},
      key: null,
      ref: null,
      props: {},
      _owner: null,
      _store: {}
    },
    {
      type: function ComponentB(){},
      key: null,
      ref: null,
      props: {
        className: 'ComponentB-style'
      },
      _owner: null,
    },
  ],
  _owner: null
}
```

看到 App 组件实例对象数据结构，这就是之前介绍的 React 虚拟 DOM 数据结构，同样也是 react 和 react-dom 最紧密关联的部分。

1.3.2 组件中的状态响应

既然负责实现 react 特性的是 react-dom，并且 React 是数据驱动视图，那么在数据更新页面重渲染的时候，React 中的 setState 是如何触发状态响应，链接到 react-dom 从而更新视图的呢？

先说结论：当调用 react 特性如 setState、Hook 修改组件状态时，其实就是异步调用了 react-dom 里的方法，而 react-dom 的方法会在创建类组件或函数组件过程中，注入 react 中。接下来开始探索向类组件和函数组件注入更新器的过程。

先说类组件，在 React 中定义了 Component 类的属性和方法，从定义中可以看到有一个 updater 实例属性，而 this.updater.enqueueSetState 就是其有效操作，用于触发 react-dom 中的页面重渲染，类组件声明源码路径位于 src/ReactBaseClassed.js 文件中，如图 1.12 所示。

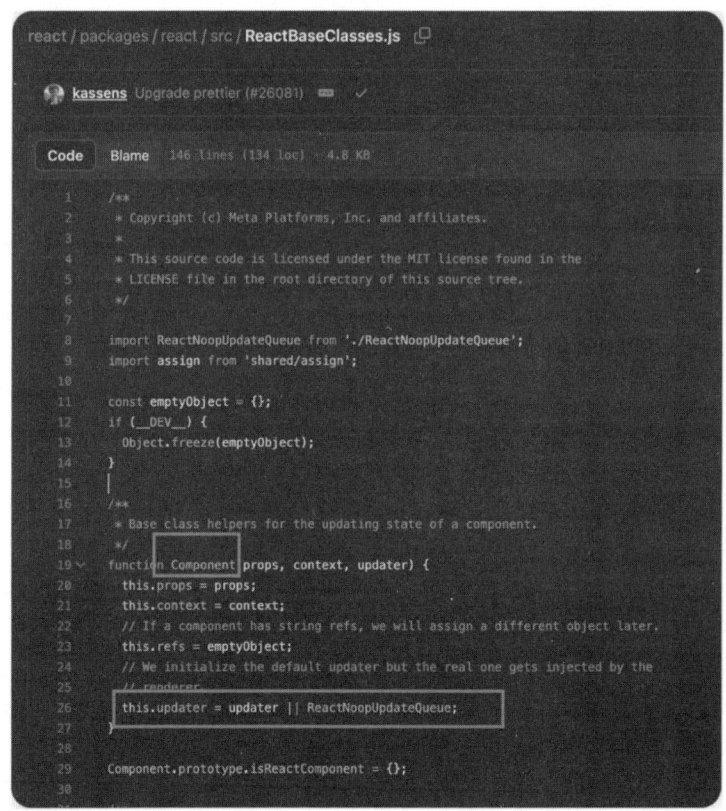

图 1.12

再说函数组件，函数组件会通过 useState 的第二个出参返回值来更新状态，本质上是调用了 useState 中的 ReactCurrentDispatcher.current.useState。

useState 的源码路径：react/src/ReactHooks.js，如图 1.13 所示。

```
react / packages / react / src / ReactHooks.js

Code   Blame   239 lines (213 loc) · 7.26 KB

export function useState<S>(
  initialState: (() => S) | S,
): [S, Dispatch<BasicStateAction<S>>] {
  const dispatcher = resolveDispatcher();
  return dispatcher.useState(initialState);
}
```

图 1.13

resolveDispatcher 函数的路径也在 ReactHooks.js 中，如图 1.14 所示。

```
react / packages / react / src / ReactHooks.js

Code   Blame   239 lines (213 loc) · 7.26 KB

function resolveDispatcher() {
  const dispatcher = ReactCurrentDispatcher.current;
  if (__DEV__) {
    if (dispatcher === null) {
      console.error(
        'Invalid hook call. Hooks can only be called inside of the body of a function component. This could happen for' +
          ' one of the following reasons:\n' +
          '1. You might have mismatching versions of React and the renderer (such as React DOM)\n' +
          '2. You might be breaking the Rules of Hooks\n' +
          '3. You might have more than one copy of React in the same app\n' +
          'See https://reactjs.org/link/invalid-hook-call for tips about how to debug and fix this problem.',
      );
    }
  }
  // Will result in a null access error if accessed outside render phase. We
  // intentionally don't throw our own error because this is in a hot path.
  // Also helps ensure this is inlined.
  return ((dispatcher: any): Dispatcher);
}
```

图 1.14

因此，我们找到了从 react 定义特性到 react-dom 中实现特性的入口。可以得出如下结论：

- 在创建类组件实例阶段，react-dom 设置 updater，setState 更新状态时会调用 updater 的 enqueueSetState 方法，从而调用 react-dom 更新视图；
- 在创建函数组件前，react-dom 覆盖了 ReactCurrentDispatcher 的 current，创建函数组件时，调用了 react-dom 中的 Hook。

1.3.3 React 基本引入方式

接下来我们一起体验 React 的几种基本引入方式。首先对于传统的前端项目，通过<script>标签引入 JS 文件从而引入 React，就像是这样：

```
<body>
  <!-- HTML 模板 -->
  <!-- 加载 React -> React 组件、ReactDOM -> DOM 相关 -->
  <!-- 注意：部署时，将"development.js"替换为"production.js" -->
  <script src=
  "https://unpkg.com/react@16/umd/react.development.js">
  </script>
  <script src=
  "https://unpkg.com/react-dom@16/umd/react-dom.development.js">
  </script>
  <!-- 引入 React 代码的文件 -->
  <script src="react_use.js"></script>
</body>
```

此时打印 window.react，你就会看到挂载的 react 对象了。然后在 HTML 中定义容器元素，也是整个页面的唯一根元素，通过使用 document.querySelector + ID 的方式来渲染，接着开始编写 react_use.js 文件。回忆一下前面讲解的虚拟 DOM，虚拟 DOM 的核心是什么？其实就是 React.createElement 函数。简单写几个元素来感受回忆一下：

```
import React from 'react';
const ReactChildComponent = React.createElement(
  'p',
  { style: { color: 'red', paddingTop: 10 } },
  '子组件'
);
const ReactComponent = React.createElement(
  'div',
  { onClick: () => this.setState({ liked: true }) },
  ReactChildComponent,
);
const domContainer = document.querySelector('#react_container');
ReactDOM.render(ReactComponent, domContainer);
```

但是这样的代码可读性很差，随着节点层级的增加，嵌套深度会线性增加。前面曾提及，一般在实际开发中会通过 JSX 的形式来提高效率和可读性，因此可以这样写：

```
const ReactComponent = (
  <div>
    <span>你好</span>
```

```
      <p>子组件</p>
    </div>
);
const domContainer = document.querySelector('#react_container');
ReactDOM.render(ReactComponent, domContainer);
```

但是，JSX 并不是浏览器原生支持的特性，因此需要使用 Babel 插件将其编译为标准的 JavaScript 语法。我们需要在 react_use.js 文件之前引入 Babel 插件：

```
<script src="https://unpkage.com/babel-standalone@6/babel.min.js">
</script>
```

给你想使用 JSX 语法的 script 标签添加 type="text/label"属性，接下来就可以在任何<script>标签中使用 JSX 语法了，但是这种方式只适合入门学习和测试（若每次刷新页面都需要将代码通过 Babel 插件编译一遍，会使页面响应变慢），不适用于生产环境。

1.3.4　开箱即用的 React 引入

1.3.3 节的方法的确可以快速"无污染"地引入 React 并使用，但其并不是最完美的方案。对此，你也可以通过一个脚手架工具来把所有环境都搭建好，包括但不限于本地开发服务器、代码包构建编译、包依赖管理器、React 环境、React 路由环境、React 状态管理库初始化等，你只需要关注代码本身而不用去关注 Babel、Webpack 这些杂乱的配置。这一工具就是 Create React App 工具。以下仅简单介绍 Create React App 的使用及其功能，具体内容建议读者参考官方文档。

首先，在你想要创建 React 应用的目录中打开终端，输入命令：

```
npx create-react-app
cd my-app
npm start
```

之后你会发现生成了一个如图 1.15 所示的目录结构的文件包。

其提供了开箱即用的启动本地 React 应用服务的能力，你可以通过 npm run start 终端命令访问应用主页面。同时，由于 Create React App 基于 Webpack 构建，其还提供了

```
my-app
├── README.md
├── node_modules
├── package.json
├── .gitignore
├── public
│   ├── favicon.ico
│   ├── index.html
│   └── manifest.json
└── src
    ├── App.css
    ├── App.js
    ├── App.test.js
    ├── index.css
    ├── index.js
    ├── logo.svg
    └── serviceWorker.js
```

图 1.15

许多基础的配置化能力，包括但不限于代码分割、生产/开发环境分环境打包等。建议在同一个工程应用下使用同一个 React 版本，可以通过 npm-lock 的形式锁定 react 和 reactDOM 的版本以避免工程中出现第三方包相关的异常。

对于已有的其他 Webpack 项目，引入 React 也很简单，就像正常的 npm 包一样使用即可，但需要通过 Babel 插件处理 JSX 语法。

1.3.5　Vite 快速初始化

在浏览器支持 ES 模块之前，JavaScript 并没有提供原生能力给开发者，尤其是当构建大型项目时，随着编码量不断增加，包含几千个模块（JS 文件）的应用也非常普遍，对于启动本地服务或是打包生产资源都需要很长时间，这极大地降低了开发者的开发体验和效率。而 Vite 的出现就是为了解决这些问题。Vite 对于 React 也提供了开箱即用的应用环境构建能力，只需要在终端输入命令：

```
npm create vite
```

执行命令后，npm 首先会自动下载第三方包 create-vite，然后执行这个包中的项目初始化逻辑，随后还会进行一系列的终端交互。如图 1.16 所示，选择想要通过 Vite 构建的应用类型即可，这里选择 React。

```
✓ Project name: vite-project
? Select a framework: › - Use arrow-keys. Return to submit.
    vanilla  // 无前端框架
    vue      // 基于 Vue
  > react    // 基于 React
    preact   // 基于 Preact（一款精简版的类 React 框架）
    lit      // 基于 lit（一款 Web Components 框架）
    svelte   // 基于 Svelte
```

图 1.16

此时已经通过 Vite 在当前目录创建好了 React + Vite 应用，紧接着和 Create React App 一样，执行以下命令启动本地项目：

```
cd vite-project
npm install
npm run dev
```

很快，我们就可以在浏览器中通过打开默认端口（如 localhost:5173）看到应用首页了，如图 1.17 所示。

观察项目结构，会发现和 Create React App 比较相似，项目的目录结构如图 1.18 所示。

和其他前端构建工具创建出来的项目一样，Vite 构建出来的项目入口也是

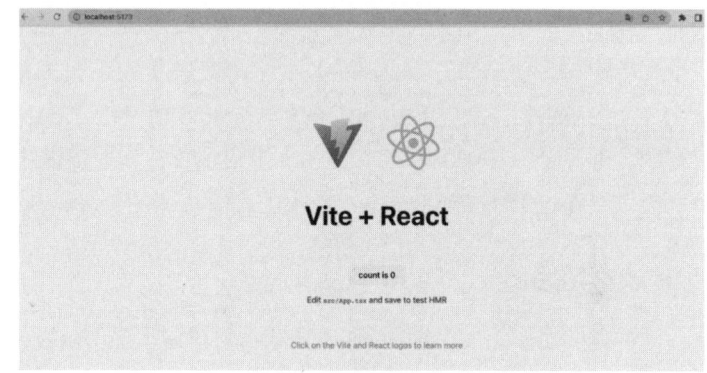

图 1.17

index.html。传统的前端项目通常由多个 HTML 页面组成，每个页面通过 CSS 和 JavaScript 来实现样式和交互。而 React 应用则通过一个入口文件启动，其内部由多个组件构成，每个组件都有自己的模板和逻辑。也就是说，当我们访问 localhost:5173 的时候，Vite 启动的本地服务器会自动返回这个 HTML 文件的内容，并且将 main.tsx、App.tsx 的内容先通过 Babel 把 React JSX 语法解析为原生 JS 语法，最后加载到 HTML 文件中去，从而生成页面。

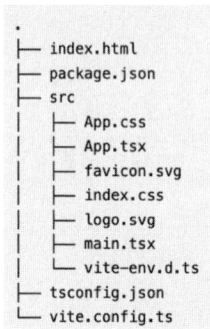

图 1.18

这里的 main.tsx 代表入口，App.tsx 代表入口组件。可以看到 index.html 中引入了 main.tsx，即入口文件引入了入口组件的依赖关系。

```
<!DOCTYPE html>
<html lang="en">
<head>
  <meta charset="UTF-8">
  <meta http-equiv="X-UA-Compatible" content="IE=edge">
  <meta name="viewport"
        content="width=device-width, initial-scale=1.0">
  <title>Vite + React + TS</title>
</head>
<body>
  <div id="root"></div>
  <script type="module" src="/src/main.tsx"></script>
</body>
</html>
```

仔细观察，body 标签简单明确，由以下两部分组成。

- id 为 root 的空 div 节点：整个 React 应用无论多么庞大，始终是以根节点为 root 的树结构不断向下衍生，以一棵树的形式描述虚拟 DOM。
- 引入 main.tsx 的原生 ES Module 脚本：从 Chrome 的某个版本开始，主流浏览器已支持原生 ES 模块规范。因此，使用原生 ES 语法的脚本可以直接在浏览器中执行，只需在 script 标签中声明 type="module"。

main.tsx 的内容如下：

```
import React from 'react';
import ReactDOM from 'react-dom/client';
import App from './App';
import './index.css';
ReactDOM.createRoot(document.getElementById('root') as HTMLElement).render(
  <React.StrictMode>
    <App />
  </React.StrictMode>,
);
```

可以看到，main.tsx 引入了 App 组件，而该项目所生成的第一个虚拟 DOM 节点其实就是 React.createElement('root', <App />)。

在开发阶段，Vite 为了加快项目的构建速度，通过 Dev Server 实现了不打包的特性；而在生产环境中，Vite 会基于 Rollup 进行打包，并且采取一系列的打包优化手段。关于这一点，可以在 package.json 中找到答案：

```
"scripts": {
  "dev": "vite",                    // 开发阶段启动 dev server
  "build": "tsc && vite build",     // 生产环境打包
  "preview": "vite preview"         // 生产环境打包后预览产物
}
```

看到生产环境打包，你可能比较疑惑，为什么在执行 vite build 之前要先进行 tsc 呢？tsc 是 TypeScript 的官方编译命令，可以通过编译 TypeScript 代码进行类型校验。Vite 底层并没有实现自己的类型系统，而是先借助了 TypeScript 的类型检查能力，再借助 Rollup 的打包构建能力来生产线上包，最后可以通过 npm run preview 命令预览打包后的应用。

思考题

1. 什么是单页面应用（SPA）？它与多页面应用有何不同？

2．简述模块化、组件化的概念与以往前端技术架构的区别以及优势？

3．虚拟 DOM 是什么？它和真实 DOM 有什么区别？它在页面中真实存在吗？

4．diff 算法和 React key 的关系是什么？key 可以用 Math.random()吗？为什么？

5．在项目中 react 和 react-dom 应该先引入哪个包？如果引入顺序错了会发生什么？

6．组件中的状态响应一次，会经历些什么过程？其和 CSS 变化导致的页面重绘有什么区别？

第 2 章
组件与视图渲染

React 组件是构建 React 应用的基本单元，它们可以是类组件或函数组件，每个组件都用于表示用户界面的一部分。类比于传统前端开发的 div、span、b 等构筑了原生 HTML，React 应用的视图的渲染由无数个 React 组件构成。

2.1 React 组件初探

组件（component）是 React 的核心概念之一，它是构建 Web 应用程序界面的基石。比如在对 UI 模块进行拆分时，可能需要将一个页面拆成很多个 div 元素，如导航栏块、侧边栏块、主体块、底部块，此时，就可以把组件当作 div 的完全替代品。

React 组件的使用方式就像是 HTML 标签一样，可以组合、排序、嵌套各种组件来构建出一个完整页面，例如下面这段代码块就是由 React 组件组成的：

```
const Title = () => { return <h1>我是标题栏</h1>;};
const Sidebar = () => { return <div>我是侧边栏</div>;};
const Content = () => { return <div>我是页面主体</div>;};
const Footer = () => { return <footer>我是页面底部</footer>;};
const App = () => {
  return (
    <>
      <Title />
      <div style={{ display: 'flex' }}>
        <Sidebar />
        <Content />
      </div>
      <Footer />
    </>
  );
};
```

上述代码呈现了组件的定义和使用方法，你应该能很快联想出这个 App 组件（应用程序根组件）所呈现的页面架构。此外，基于 React 组件的特性，也可以把 Title、Sidebar、Content、Footer 组件抽离到公共组件目录中，从而在项目中的多个页面去使用。这就是组件的复用性。

通常，每个组件会单独抽离成一个文件，并且会把组件向外导出。export default 作为标准 JavaScript 语法（非 React 专用）用于导出所定义的 React 组件，并通过 import from 形式导入所定义的组件。下面的代码就很好地表达了 React 通过模块化实现一个组件的导入和导出。

定义与导出：

```
export default function Component() {
  return <div>我是 Component 组件</div>;
}
```

导入：

```
import Component from './Component';
```

2.1.1 类组件

类组件在 React v16 之前是主流方式，但目前 React 已经更新至 v19 版本，函数组件已成为 React 的主流组件编程范式，因此本小节仅对类组件做简要的介绍。

类组件的定义有如下要求：

- 类组件必须继承于 React.Component；
- 类组件必须声明实现 render 函数。

在 ES6 之前，可以通过 create-react-class 模块来定义类组件，但笔者还是建议使用官网案例，即使用类定义一个组件：

```
class App extends React.Component {
  constructor() {
    super();
    this.state = {};
  }
  render() {
    return <h2>我是类组件</h2>;
  }
};
```

可见类组件由 3 部分构成：

- constructor：类组件的构造函数，为可选函数，通常在 constructor 中初始化组件的状态；

- this.state：组件中的状态（数据），用于维护组件内部的动态数据，可以通过 this.state.<属性名>来访问某个状态；
- render()：类组件中唯一必须声明实现的方法，表现了组件模板（也可以由 state 组成）。

关于 state，可以在函数中或在模板中通过 this.state 获取和展示状态的某个属性值，就像这样：

```
class App extends React.Component {
  constructor() {
    super();
    this.state = {
      name: 'React Component',
    };
  }

  click() {
    console.log(this.state.name);
  }

  render() {
    return <h2 onClick={this.click}>我是类{this.state.name}</h2>;
  }
};
```

那么应该如何修改组件中的状态呢？比如点击某个元素，改变 this.state.name，可以通过 setState()来更新数据，更新后 React 会帮我们执行 render()去重渲染一次页面，将页面中用到的 state 中的数据全部更新，就像这样：

```
class App extends React.Component {
  constructor() {
    super();
    this.state = {
      name: 'React Component',
    };
  }

  click() {
    this.setState({
      name: 'React Component update!'
    });
  }

  render() {
    return <h2 onClick={this.click}>我是类{this.state.name}</h2>;
```

```
    }
};
```

对于 render 函数,在页面渲染时,会默认调用一次 render,展示初始的状态值(React Component),也被称为组件首次渲染;当 state 改变后,组件依赖于 state,React 会监听到 state 变化,进行一次重渲染,在页面中展现新数据。具体的操作参考虚拟 DOM 与 diff 算法相关内容。

2.1.2 函数组件

顾名思义,函数组件就是使用 function 来定义的组件,以函数返回值对应类件的 render 方法。与类组件相比,函数组件有自己的特点:

- 写法更自由灵活,函数式组件依赖于 React 的 Hook 来实现各种能力,如组件生命周期、state(类组件依赖于 React.Component 的上层类实现 state、依赖于 React 的特定生命周期钩子函数,如 componentDidMount 函数)。
- 函数组件没有 this 组件实例。

声明一个函数组件就像编写 JavaScript 函数一样的自然,就像这样:

```
const App = () => {
  return <h2>我是函数式组件</h2>
};
```

如果是带有组件状态和点击交互逻辑的函数式组件,则需要借助 Hook 来声明 state,代码块如下:

```
const App = () => {
  const [name, setName] = React.useState('React Component');
  const click = () => {
    setName('React Component update!');
  };
  return <h2 onClick={click}>我是函数{name}</h2>;
};
```

同样的需求,对比类组件,函数组件会显得更简洁直观。

2.2 组件状态与通信

上一节介绍了基本的 React 组件使用方式,以及状态(state)如何驱动视图更新的概念,为了更深入地理解 React 的工作原理,本节将进一步探讨一些核心概念,包括 state、props 和组件之间的通信机制。这些概念将帮助我们更好地掌握如何构建灵活且可维护的 React 应用。

2.2.1　state

React 把组件看成一个状态机（state machines），通过与用户的交互，实现不同状态，最终渲染用户界面，让用户的视图层和组件中声明的数据层保持一致。在 React 中，只需要更新组件的 state，React 会自动根据新的 state（虚拟 DOM 树）来重新渲染用户界面，因此在大部分情况下不需要去获取和操作 DOM。

我们可以在组件初始化时进行 state 声明。对于类组件可以像这样去声明：

```
export default class Index extends React.Component{
  constructor(){
    super();
    this.state = {
      name:'小明'
    }
  }
  render(){
    return (
      {this.state.name}
    )
  }
};
```

通过 this.state 给组件设定一个初始的 state，当第一次 render 时就会通过此数据来渲染组件，并且 state 不同于 props（不可变），state 是可以被改变的，并且改变后会引入页面重渲染。需要注意的是，我们只能通过 this.state=来初始化，而在 constructor 中是唯一能够初始化的地方。不过，我们无法通过 "this.state=" 的方式来修改 state，而需要通过 this.setState()方法来修改。

比如，通过异步操作获取数据。让我们模拟一个页面渲染调用网络请求获取数据后更新状态的场景：

```
componentDidMount(){
  fetch('url')
    .then(response => response.json())
    .then((data) => {
      this.setState({name:data});
    }
};
```

componentDidMount 是 React 中类组件自带的生命周期钩子函数，表示组件初始化渲染完毕的时期回调。当调用 this.setState 方法后，React 会更新组件内部 state，并且再次调用 render 方法，对组件进行重新渲染。

还记得之前提过的批量更新只进行一次 DOM 操作吗？看到这里，或许你会疑惑：如果在一段代码块中多次调用 this.setState，难道是调用几次就执行几次 DOM 操作吗？答案是否定的，在 React 底层中如果多次调用会将状态更改的任务放入一个异步队列中，当一段代码块运行完后进行批量更新。

如果有这样一段代码块：

```
componentDidMount(){
  fetch('url')
    .then(response => response.json())
    .then((data) => {
      this.setState('我是 A');
      this.setState('我是 B');
      this.setState({name:data});
    }
};
```

在得到接口返回的数据后，其执行两次 this.setState，而 React 最终只会执行一次更新，也就是最后一次，由于对同一个状态进行更新，React 只会取最后一次的值作为新值。

另外，setState 还可以接收第二个参数，它是一个函数，会在 setState 调用完成并且组件开始重新渲染时被调用，可以用来监听状态更新是否完成。

```
this.setState({name:data}, () => {
  console.log('渲染完成，状态已更新')
});
```

对于函数式组件，底层的状态更新原理其实是一样的，只是写法上会有一些区别。可以使用 React.useState 来声明一组单个的状态和更新其状态的函数：

```
const Index = () => {
  const [name, setName] = React.useState('小明');
  return (
    {name}
  )
};
```

对于页面加载调用接口得到响应值后更新状态的场景，在函数式组件中的写法是这样的：

```
const Index = () => {
```

```
  const [name, setName] = React.useState('小明');
  React.useEffect(() => {
    fetch('url')
      .then(response => response.json())
      .then((data) => {
        setName(data);
      });
  }, []);
  return (
    {name}
  )
};
```

在接口响应时,调用 React.useState 返回的第二个参数来更新状态,并且对于 componentDidMount 生命周期钩子函数,在函数式组件中也有了新的替代,这部分内容在后面关于 Hook 的章节会详细讲解。

讲完了 React state 的基本使用方式后,我们来梳理一下在 react 源码视角中是如何实现组件更新的。对于整个调用的一次 setState 更新状态,可拆解为四步:

① 当 this.setState 方法被调用,会将更新的状态传递给组件更新器(updater),让组件更新器临时存储新状态。每个组件都会有自己唯一的 updater 更新器实例,当组件更新时调用 updater.addState(newState)。

② 状态临时保存完成后会判断是否为批量更新模式,如果是,则将组件更新器放入更新队列中;如果不是,则直接更新组件。

③ 更新组件时,需要判断此时是否有状态需要更新,如果有则先计算更新状态,将二次更新的状态再重新设置给组件。

④ 组件状态计算完毕后,再调用 React.render 方法获取新的虚拟 DOM,进行 diff 对比,最后将差异性更新到真实 DOM 对象中。

具体的 react 实现源码如下:

```
// 更新器队列
const updateQueue = {
  // 是否处于批量更新模式
  isBatchingUpdate: false,
  // 存储更新器
  updaters: [],
  // 添加更新器
  add(updater) {
    this.updaters.push(updater);
  },
  // 批量更新
```

```
  batchUpdate() {
    let updater;
    while((updater = this.updaters.shift())) {
      updater.updateComponent()
    }
    this.isBatchingUpdate = false
  }
};

// 组件更新器
class Updater {
  constructor(instance) {
    // 存储组件实例对象，当状态处理完成用于更新组件
    this.instance = instance;
    // 缓冲队列，在判断批量和非批量之前存储状态
    this.pendingState = [];
  }
  addState(newState) {
    // 将状态放入缓冲队列
    this.pendingState.push(newState);
    //判断是批量更新模式还是非批量更新模式
    //如果是批量更新模式，则直接更新组件
    //如果是非批量更新模式，则添加到更新队列，等待
    updateQueue.isBatchingUpdate ? updateQueue.add(this) : this.updateComponent();
  };
  // 更新组件的方法
  updateComponent() {
    let { instance, pendingState } = this;
    // 如果有状态需要更新
    if(pendingState.length > 0) {
      // 重新设置组件状态
      instance.state = this.getState();
      // 重新渲染组件
      instance.forceUpdate();
    }
  };
  // 重新设置组件状态
  getState(){
    // 获取组件实例和待更新状态
    let {instance, pendingState} = this;
    // 获取当前状态
```

```
    let {state} = instance;
    // 获取待更新状态
    let nextState = state;
    // 遍历所有状态，合并操作
    pendingState.forEach(partialState => {
      // 如果当前是函数状态
      if(isFunction(partialState)) {
          // 调用函数返回新状态
          nextState = partialState(state);
      } else {
          // 如果是对象，合并状态
          nextState = {...nextState, ...partialState}
      }
    })
    pendingState.length = 0;;
    return nextState;
  }
};

// React.Component 类组件继承父类
class Component {
  constructor() {
    // 组件状态对象，对应类组件中的状态初始化声明
    this.state = {};
    // 组件更新器
    this.$updater = new Updater(this);
  }
  setState(newState) {
    // 更新状态，首先给到更新器缓存队列中
    this.$updater.addState(newState);
  }
  forceUpdate() {
    // 生成新虚拟 DOM
    const newVdom = this.render();
    // 获取旧 DOM
    const dom = this.dom();
    // 根据旧 DOM 生成对应虚拟 DOM
    const oldVdom = dom._vistualDOM;
    // 进行新旧虚拟 DOM 对比
    diff(newVdom, oldVdom, dom, container);
  }
};
```

从代码块中可以看到，整体更新链路分为两条：第一条是批量更新，会把更新器放入更新队列中一次性更新；第二条则是非批量更新（同步更新），有新增的状态变化就立即执行状态更新并进行 DOM diff 操作更新到视图中。这就是 react 中 setState 的核心逻辑。

2.2.2 props

props 是属性的简写，是用于在组件之间传递数据的重要机制，通过它可以在父组件和子组件之间建立数据联系，实现组件的复用和动态数据更新。如父组件可以在其 state 中定义值，并通过 props 将这些值传递给子组件。

props 的核心更新链路是通过父组件改变 props 源来触发子组件的更新。其与 state 的更新相比，主要有以下几点区别：

● state 是可读可写的，可用于用户进行交互修改页面状态，而 props 是只读的；

● state 主要用于控制组件自身，而 props 主要用于组件本身的状态传递，使用最广泛的场景就是父组件把自身 state 作为 props 传递给子组件，子组件作为消费方不可修改 props，因此 props 是一种单向的数据传递方式。

在设计公共组件（common component）时，就像设计一个 utils 工具函数或模块一样，我们常希望组件能够复用，在这种场景下就需要使用到 props。props 就像是工具函数的入参，可以动态地将数据传递到公共组件中，提高组件的灵活性。

props 最重要的作用就是子组件间自上而下的信息传递，让我们来看一个 props 的示例：

```
const Component = (props) => {
  const { number } = props;
  return (
    <div>我是序号{number}子组件</div>
  )
};
const App = () => {
  return (
    <div>
      <Component number={1} />
      <Component number={2} />
      <Component number={3} />
    </div>
  )
};
```

App 通过 Component 的 number 属性进行动态传递，在渲染时代码灵活度就会提高很多。那如果是结合日常业务场景的列表渲染呢？就像图 2.1 所示。

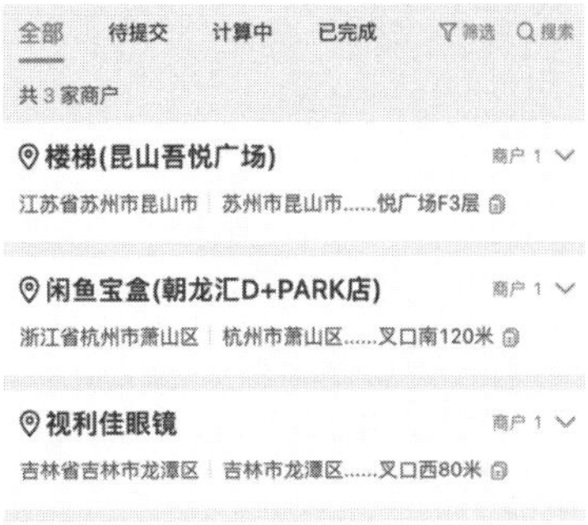

对于业务中常见的列表渲染结合 props 子组件动态传参，不难想到，可以将每一个卡片封装成一个 Card 组件，在父组件中进行数据请求，当获取到接口响应数据后遍历渲染数组长度个数的 Card 组件，就像这样：

图 2.1

```
import { useEffect, useState } from 'react';
const Card = (props) => {
  const { data } = props;
  return (
    <div>
      <h1>{data.title}</h1>
      <p>{data.desc}</p>
      <footer>{data.footer}</footer>
    </div>
  );
};

const App = () => {
  const [data, setData] = useState([]);
  useEffect(() => {
    getCardInfo();
  }, []);
  const getCardInfo = () => {
    fetch('http://example.com/movies.json')
      .then((response) => response.json())
      .then((res) => {
        if (res.data && res.success) {
          setData(res.data);
        }
      });
  };
  return (
    <div>
      {data.map((item) => {
        return <Card data={item} />;
      })}
```

第 2 章 组件与视图渲染　037

```
      </div>
    );
  };
```

如果准备使用 React 在工作中开发组件，上述这段代码是再常见不过的了，它有以下几个关键点：

- 在父组件中进行数据的获取、控制，以及对子组件的控制；
- 子组件只进行数据的渲染，业务逻辑性主要都在父组件中；
- useEffect 中进行数据获取，useState 中声明数据，由父组件决定页面重渲染。

核心业务都在父组件中，包括数据的获取、传递以及数据获取的触发时机。如果再进一步扩展，比如加一个分页器来加强父子组件的交互性，就像是这样：

```
const App = () => {
  const [data, setData] = useState([]);
  const pageCurrent = useRef({
    pageNum: 1,
    pageSize: 10
  })
  useEffect(() => {
    getCardInfo();
  }, []);

  /**
   * @description: 获取列表数据
   * @return {*}
   */
  const getCardInfo = () => {
    fetch('http://example.com/movies.json', {
      pageNum: pageCurrent.pageNum,
      pageSize: pageCurrent.pageSize
    })
    .then((response) => response.json())
    .then((res) => {
      if (res.data && res.success) {
        setData(res.data);
      }
    });
  };

  /**
   * @description: 下一页
   * @return {*}
```

```
*/
const toNextPage = () => {
  const { pageNum, pageSize } = pageCurrent.current;
  pageCurrent.current = {pageNum: pageNum + 1,pageSize};
  getCardInfo();
};

return (
  <div className={styles['container']}>
    <div className={styles['list']}>
      {data.map((item) => {
        return <Card data={item} />;
      })}
    </div>
    <div onClick={toNextPage}>下一页</div>
  </div>
);
};
```

在这段代码中,相比于之前一段代码增加了分页的交互,当点击下一页,会重新获取列表中的数据,再一次调用接口,当接口返回数据后,所有的 Card 组件的 props 也重新刷新,父组件导致子组件重新渲染。

如果有一个场景,每个 Card 组件中有一个更新的按钮,当点击更新,子组件发起一个重新获取列表数据的执行请求给父组件,父组件监听后进行列表刷新,这种情况应该怎么做呢? 其实很简单,只需要父组件传给子组件一个函数,当子组件需要更新数据时执行 props 中的回调函数即可,就像这样:

```
import { useEffect, useState } from 'react';
const Card = (props) => {
  const { data, onRefresh } = props;
  return (
    <div>
      <h1>{data.title}</h1>
      <p>{data.desc}</p>
      <footer>{data.footer}</footer>
      <button onClick={onRefresh}>刷新列表</button>
    </div>
  );
};
const App = () => {
  const [data, setData] = useState([]);
  useEffect(() => {
```

```
      getCardInfo();
    }, []);
    const getCardInfo = () => {
      fetch('http://example.com/movies.json')
      .then((response) => response.json())
      .then((res) => {
        if (res.data && res.success) {
          setData(res.data);
        }
      });
    };

    return (
      <div>
        {data.map((item) => {
          return <Card data={item} onRefresh={getCardInfo} />;
        })}
      </div>
    );
};
```

每个 Card 组件中都可以有一个更新按钮，若点击按钮则执行父组件传来的函数进行子传父的回调，父组件重新获取接口数据，重新渲染列表（所有的 Card 组件），导致子组件全部执行重渲染的流程。

2.2.3 props 导致的更新

父组件传给子组件的 props 更新会导致子组件更新，那么如果传给子组件的状态没更新，只是父组件的其他部分改变了，子组件还会重渲染吗？

在探究原理之前，先回忆一下 React 中的 diff 算法，它会将更新前后的两棵虚拟 DOM 树进行对比，但这并不会决定组件是否更新，只会决定是否要复用旧的节点。

举个简单的例子：

```
import { useState } from 'react';
const Child = () => {
  console.log('child render');
    return null;
};
const App = () => {
  const [name, setName] = useState(1);
```

```
  return (
    <div onClick={() => setName(2)}>
      <Child />
    </div>
  );
};
```

Child 组件没有接收来自父组件的值，每次点击父组件元素让 name 更新，Child 组件会更新吗？答案是会的。你一定会好奇，子组件没有接收任何的 props，为什么也会更新呢？

首先，父组件经过了 diff 阶段，会判断 Child 组件是否发生变化，在本案例中 Child 内部的元素结构和状态无任何变化。React 还会对比 Child 组件前后的 props 是否相同，在本案例中，前后 props 不相同。

对此，你一定忍不住要问，明明没传 props，为什么不相同？原因是 React 内部对于 props 的对比只进行了浅层比较，通过 !== 来判断，这样即使没传 props，每次生成的 props 对象也都是新的指针，即使为空，也会生成不同的 props 空对象，就像这样：

```
const oldProps = current.memoizedProps; //更新前旧的 props
const newProps = workInProgress.pendingProps; //待比较更新的 props
if (oldProps !== newProps) {
    didReceiveUpdate = true; //标记为发生变化，需要更新
};
```

那么，怎样才可以避免这样的无效更新呢？一共有三种方案：

方案 1：使用 React.memo，可以指定在 diff 时对于被 memo 包裹的组件只做浅层比较；

方案 2：使用 React.useMemo 或 React.useCallback 来包装子组件，让每次更新子组件都为同一个 JSX 对象，这样 props 的比较就会相同；

方案 3：将子组件作为 children 来传递。

对于方案 1，React.memo 的原理其实来源于源码中的 shallowEqual 函数，该函数会接收两个对象，分别对应旧的 props 和新的 props，一共有四种比较策略，如果四种策略都通过，则判定新旧 props 为同一个对象，不做更新，复用旧的节点。四种策略如下：

- 判断两者是否为同一对象，不是同一对象则返回 false；
- 判断两者的值若不为 object 或为 null，则返回 false；
- 对比两者 key 的数量，不一致则返回 false；
- 对比两者 key 的值是否相同，不一致则返回 false。

源码如下:

```
// shallowEqual.js
function shallowEqual(objA: mixed, objB: mixed): boolean {
  // 一样的对象返回 true
  if (Object.is(objA, objB)) {
    return true;
  }

  // 若不是同一对象或者为 null 返回 false
  if (
    typeof objA !== 'object' || objA === null ||
    typeof objB !== 'object' || objB === null
  ) {
    return false;
  }
  const keysA = Object.keys(objA);
  const keysB = Object.keys(objB);

  // 若 key 数量不同，则返回 false
  if (keysA.length !== keysB.length) {
      return false;
  }

  //对比 key 的值，不相同则返回 false
  for (let i = 0; i < keysA.length; i++) {
    if (
      !hasOwnProperty.call(objB, keysA[i]) ||
      !Object.is(objA[keysA[i]], objB[keysA[i]])
    ) {
      return false;
    }
  }
  return true;
}
```

可以看到浅比较 props 的实现原理很简单，对应着上述四种策略。

对于方案 2，如果不了解 React.useMemo 和 React.useCallback，没有关系，先看一下这段代码块：

```
import { useMemo } from 'react';
const Child = () => {
  console.log('child render');
```

```
  return null;
};
const App = () => {
  const [name, setName] = useState(1);
  const child = useMemo(() => <Child />, []);
  return <div onClick={() => setName(2)}>{child}</div>;
};
```

React.useMemo 接收两个参数,第一个参数为返回值,第二个参数为依赖项,当依赖项数组中的值发生变化时,返回值会重新计算,也就是说若第二个依赖项传空数组,则依赖项永远都不会发生变化,Child 组件经过 React.useMemo 包裹后就一直不会被 React 去计算 diff,如此就实现了父组件更新,而子组件不触发更新。

但对于 React.useMemo 的使用,如果传给了子组件值,但是未声明依赖项,就会导致子组件一直不发生变化,就像这样:

```
import { useMemo } from 'react';
const Child = ({name}) => {
  console.log('child render');
  return name;
};
const App = () => {
  const [name, setName] = useState(1);
  const child = useMemo(() => <Child name={name} />, []);
  return <div onClick={() => setName(2)}>{child}</div>;
};
```

像这种情况,父组件将 name 传给了子组件,但是由于子组件未声明 name 为改变依赖项,因此当 name 发生变化时,子组件依然会返回初始值 1。因此,对于 React.useMemo 的缓存策略,在优化时也需要充分考虑可能发生的"意外"。

对于方案 3,可以简单理解成向上提升和向下移动 state。先看一个案例:

```
const App = () => {
  const [color, setColor] = useState('red');
  return (
    <div>
      <input value={color} onChange={(e) => setColor(e.target.value)} />
      <p style={{ color }}>Hello, world!</p>
      <Child />
    </div>
  );
};
```

input 的 onChange 事件是一个频繁触发的颜色指示器，每秒可触发上百次，而 Child 组件是一个固定渲染且不依赖父组件状态的子组件，如何通过状态向下移动的方式来避免 Child 组件被渲染呢？只需要将下面这段性能消耗严重的代码抽离到单独的一个 Form 组件中，同时把 color 状态单独交给 Form 组件去管理，这样 App 父组件一直没有发生重渲染，Child 子组件也不会被影响，只有 Form 子组件在单独发生交互，这种方案更像是一个状态下移和隔离。

```
const App = () => {
  return (
    <div>
      <Form />
      <Child />
    </div>
  );
};
```

还有一种解法就是状态提升。可以把这段性能消耗严重的代码同样单独封装成一个组件，将 Child 子组件的内容传递给 Form 子组件，就像这样：

```
const Form = ({ children }) => {
  const [color, setColor] = useState('red');
  return (
    <div style={{ color }}>
      <input value={color} onChange={(e) => setColor(e.target.value)} />
      {children}
    </div>
  );
};
const App = () => {
  return (
    <div>
      <Form>
        <p>Hello, world</p>
        <Child />
      </Form>
    </div>
  );
};
```

其实该思路与组件整体抽离从而解耦状态是一样的，即把性能消耗严重的一部分单独抽离到一个组件中，将相对不期望被影响的一部分通过特定形式渲染，因此 Child 子组件在这种情况也不会被重新渲染。

2.2.4 父传子通信

props 是父组件传递给子组件通信方式的主要手段,就像这样:

```
const Child = (props) => {
  const { name } = props;
  return { name };
};
const App = () => {
  const [name, setName] = useState('小明');
  return <Child name={name} />;
};
```

通过 props 父传子通信的流程图如图 2.2 所示。

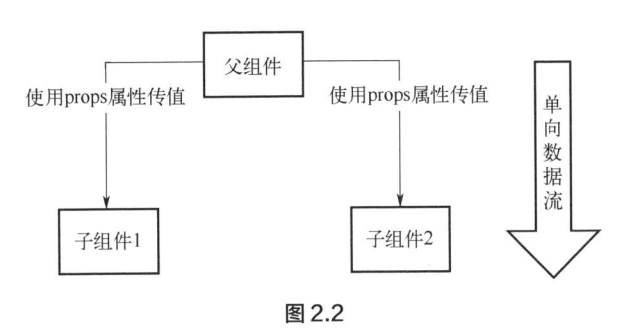

图 2.2

但如果父组件想要操作子组件的方法或者获取子组件的状态呢?想象一个场景:你有一个商品页面,商品卡片都需要定时更新自己的数据,商品页面(Shop 组件)作为父组件,商品卡片(Card 组件)作为子组件,映射成代码块是这样的:

```
import { useEffect, useState } from 'react';
const Card = () => {
  const [shopInfo, setShopInfo] = useState();
  useEffect(() => {
    getShopInfo();
  }, []);
  const getShopInfo = () => {
    fetch('url')
    .then((response) => {
      return response.json();
    })
    .then((res) => {
      setShopInfo(res);
    });
```

```
    };
  };
const Shop = () => {
  const refreshCardComponent = () => {};

  return (
    <>
      <h1>商品页面</h1>
      <button onClick={refreshCardComponent}>更新商品信息</button>
      <Card />
    </>
  );
};
```

当点击父组件中的更新商品信息时，由于 shopInfo 的数据口径由子组件自身维护，但是又暴露给父组件更新数据的触发动作，那么 refreshCardComponent 该怎么实现呢？与 props 无关，这里需要使用到 React 的三个 API：useImperativeHandle、forwordRef 和 useRef。

先看一下改造后的代码：

```
import { useEffect, useState, useImperativeHandle, forwardRef, useRef } from 'react';
const Card = forwardRef((props, ref) => {
  const [shopInfo, setShopInfo] = useState();
  useEffect(() => {
    getShopInfo();
  }, []);
    useImperativeHandle(ref, () => ({
    setShopInfo,
  }));

  const getShopInfo = () => {
    fetch('url')
    .then((response) => {
      return response.json();
    })
    .then((res) => {
      setShopInfo(res);
    });
  };
});
```

```
const Shop = () => {
  const ref = useRef();
  const refreshCardComponent = () => {
    ref.current.getShopInfo();
  };

  return (
    <>
      <h1>商品页面</h1>
      <div onClick={refreshCardComponent}>更新商品信息</div>
      <Card ref={ref} />
    </>
  );
};
```

上述代码首先在父组件中声明了一个 ref 传递给子组件，子组件通过 forwardRef 包裹组件，使得子组件内部实例可以被暴露出去，并且通过 useImperativeHandle 在子组件中把父组件需要调用的 API 暴露出去，这样父组件只需要通过传过去的 ref 来调用 getShopInfo 即可。

useRef 是 React 16.8 新增的一个 Hook API，系统性讲解参见第 4 章，本节只针对 useRef 做必要讲解。

举个例子：

```
import { useRef } from 'react';
const ref = useRef(0);
ref.current === 0; //true
```

● 返回一个可变的 ref 对象，该对象只有一个 current 属性，初始值为传入的参数，对应示例中的 0；

● 返回的 ref 对象在组件的整个生命周期内保持不变，当组件通过 state 变更时，造成的重渲染不会使 ref 值重置为初始值，可以理解为组件中的变量；

● 当更新 current 值时不会造成重渲染，页面不会发生数据更新，这是与 state 最不同的地方。

通俗地讲，useRef 就像是可以在.current 属性中保存一个可变值的"盒子"，并且 useRef 可以存储任意的变量或 DOM 节点，可以通过 useRef 声明返回的 ref 挂载在 React JSX 上的元素控制 DOM 节点，就像这样：

```
import { useRef } from 'react';
const App = () => {
  const inputRef = useRef();
```

```
  const handleFocus = () => {inputRef.current.focus();};
  return (
    <div>
      <input ref={inputRef} type="text" />
      <button onClick={handleFocus}>点击按钮聚焦输入框</button>
    </div>
  );
};
```

或者是像这样：

```
import { useState, useRef } from 'react';
const App = () => {
  const [now, setNow] = useState(Date.now());
  const ref = useRef();
  const handleStart = () => {
    ref.current = setInterval(() => {
      setNow(Date.now());
    }, 1000);
  };
  const handleStop = () => {clearInterval(ref.current);};
  return (
    <>
      <h1>Now Time : {now}</h1>
      <button onClick={handleStart}>Start</button>
      <button onClick={handleStop}>Stop</button>
    </>
  );
};
```

通过 ref 存储 setInterval 返回的 ID，需要清除时，我们只需要清除 clearInterval (ref.current)即可，这样保证了定时器 ID 的唯一性。

通常，当你的组件需要"跳出"React 并需要去与外部 API 通信时，就需要用到 ref。还有一些适用于 ref 的使用场景：

- 存储定时器 ID；
- 存储和操作 DOM 元素；
- 存储不需要计算 JSX 的其他对象。

另外，如果组件需要存储一些值，但这些值通常不会影响到 render 的逻辑，此时可以选择 ref 来存储。

接下来，什么是 forwardRef 呢？先说一下作用吧，之前说到父组件可以调用子组件暴露的能力，子组件的 ref 就是中间的桥梁，而 forwardRef 可以使得父组件得到子组件中的 DOM 节点。对于 ref 转发，React 官网中的描述是这样的：ref 转发是一项将 ref 自动地通过组件传递到其子组件的技巧。对于大多数应用的组件来说，这通常不是必需的。但对于某些组件，尤其是可重用组件库的组件则是非常有用的。

当然，如果不通过包装的 React 函数式组件，是无法直接获取到 ref 实例的，当你尝试直接传递 ref 时会报错，如图 2.3 所示。

图 2.3

此外，你也无法通过 props 来传递 ref。React 官方也表述了 ref 的使用条件，如图 2.4 所示。

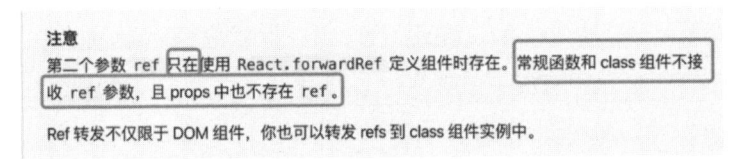

图 2.4

而 forwardRef 存在的意义就是为了解决以上问题。React.forwardRef(render)的返回值是 react 组件，入参为一个 render 函数，参数是 props 和 ref，将 ref 参数转发到组件中。对于以下两种场景很有用：
- 转发 ref 到组件内部的 DOM 节点；
- 在高阶组件中转发 ref。

举个例子：

```
import { useRef, forwardRef } from 'react';
const Child = forwardRef((props, ref) => {
  return (
    <>
      <span>我是Child组件</span>
      <input type="text" ref={ref} />
    </>
  );
});
const App = () => {
  const ref = useRef();
```

```
  const focusChildInput = () => {
    ref.current.focus();
  };
  return (
    <>
      <span>我是父组件</span>
      <button onClick={focusChildInput}>聚焦子组件输入框</button>
    </>
  );
};
```

看完上述代码不难发现 ref 转发的流程：

- 父组件创建 ref；
- 将其挂载到组件上，这个组件通过 React.forwardRef 创建；
- Child 组件接收到了一个 ref，将其转发到了 input DOM 节点；
- 现在父组件 ref.current 保存着对 input 的引用；
- 在父组件点击按钮，获取到 Child 组件内部的 input value 和原生属性。

接下来是 React 源码中 forwardRef 的实现，位于 react 项目/packages/react/src/ReactForwardRef.js 中，代码也很简单：

```
// 简化后
const REACT_FORWARD_REF_TYPE = Symbol.for('react.forward_ref');
export function forwardRef(render) {
  const elementType = {
    $$typeof: REACT_FORWARD_REF_TYPE,
    render,
  };
  return elementType;
}
```

需要注意这里的$$typeof，尽管这里显示的是 REACT_FORWARD_REF_TYPE，但最终创建的 React 元素的$$typeof 依然是 REACT_ELEMENT_TYPE。当把 forwardRef 组件转换成 ReactElement 的时候，子组件对象的 type 属性是这样的：

```
const element = {
  ?typeof: REACT_ELEMENT_TYPE,
  type: {
    ?typeof: REACT_FORWARD_REF_TYPE,
    render,
  },
};
```

那么，最后一个 API——useImperativeHandle 呢？forwardRef 可以向外暴露组件实例，useImperativeHandle 则可以理解为基于 forwardRef 的暴露后，额外自定义向外暴露组件内部的内容，如只暴露必要的方法或属性，隐藏内部实现细节。

在 React 源码中，useImperativeHandle 的执行实际上调用了 mountImperativeHandle 函数，代码如下：

```
function mountImperativeHandle<T>(
  ref: {|current: T | null|} | ((inst: T | null) => mixed) | null | voi d,
  create: () => T,
  deps: Array<mixed> | void | null,
): void {
  return mountEffectImpl(
    fiberFlags,
    HookLayout,
    imperativeHandleEffect.bind(null, create, ref),
    effectDeps,
  );
}
```

可以看到在 mountImperativeHandle 函数中又继续调用了 mountEffectImpl 方法，这个函数其实就是 useEffect 的通用函数。重点看一下 imperativeHandleEffect 的实现：

```
function imperativeHandleEffect<T>(
  create: () => T,
  ref: {| current: T | null |} | ((inst: T | null) => mixed) | null | void
) {
  if (typeof ref === "function") {
    const refCallback = ref;
    const inst = create();
    refCallback(inst);
    return () => { refCallback(null);};
  } else if (ref !== null && ref !== undefined) {
    const refObject = ref;
    const inst = create();
    refObject.current = inst;
    return () => { refObject.current = null;};
  }
}
```

在后续了解到 useEffect 和 useLayoutEffect 的设计原理时，你就会发现，useImperativeHandle 实际上就是在此基础上做了 ref 的操作，因此它并不神奇，只

第 2 章 组件与视图渲染　051

是包装了 ref 的设置和清理，与 useLayoutEffect 处于同一阶段，比 useEffect 执行时机更早。

2.2.5 子传父通信

回调函数是最常见且推荐的子传父通信方式。父组件通过 props 将回调函数传递给子组件，子组件在需要时调用这个回调函数，便可将数据作为参数传递给父组件。如图 2.5 所示。

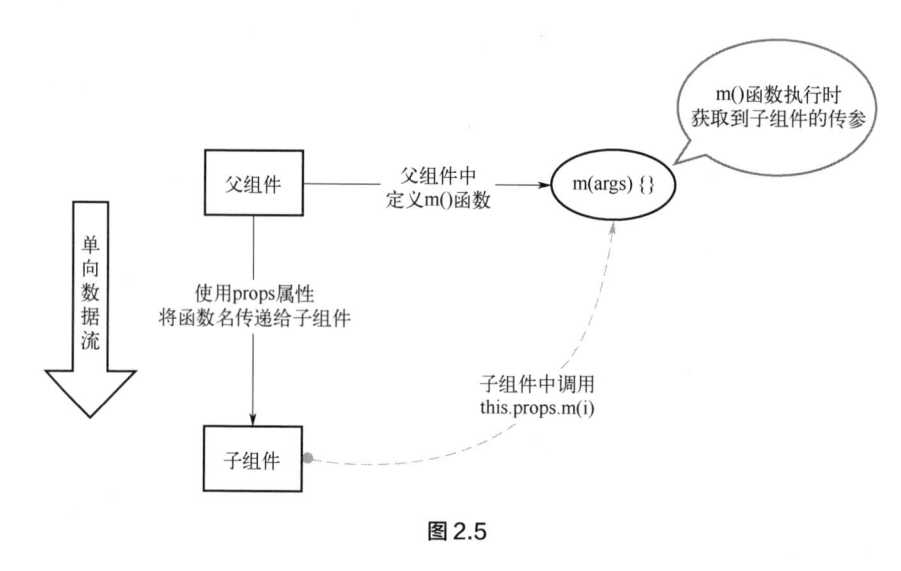

图 2.5

演示代码如下：

```
const Child = (props) => {
  const { callback } = props;
  const onClick = () => {
    callback && callback();
  };
  return (
    <>
      <span>我是子组件</span>
      <button onClick={onClick}>点我回调父组件</button>
    </>
  );
};

const App = () => {
  const callback = () => {
```

```
      console.log('父组件收到子组件的消息了');
    };
    return (
      <>
        <span>我是父组件</span>
        <Child callback={callback} />
      </>
    );
};
```

上段代码声明了 Child 子组件和 App 父组件，App 传递给 Child 一个 callback 方法。当点击 Child 组件中的按钮时，子组件会调用 callback 方法触发回调函数，便可让父组件得知回调函数被调用，进而在控制台打印出"父组件收到子组件的消息了"。

2.2.6　兄弟组件通信

同一个父组件下的两个子组件互相之间该如何通信呢？可以将它们的父组件作为一个中转站来处理，流程为：组件 A → 回调 → 父组件 → props 传值 → 组件 B，如图 2.6 所示。

图 2.6

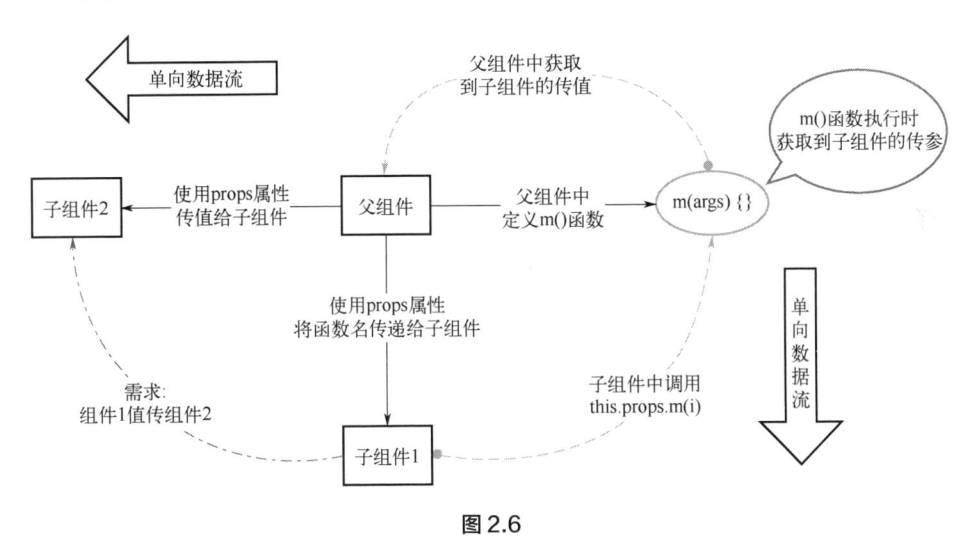

或者，也可以理解为把 2.2.4 和 2.2.5 两小节的步骤串联在一起去做，即先执行子组件传值给父组件，再执行父组件传值给子组件，代码示例如下：

```
import { useState } from 'react';
const Child1 = (props) => {
```

```jsx
  const { child1Val, callback } = props;
  const onClick = () => {
    const val = Math.random() * 1000;
    callback && callback(val);
  };
  return (
    <>
      <span>我是子组件 1</span>
      <span>{child1Val}</span>
      <button onClick={onClick}>点我回调父组件改变 Child2 的值</button>
    </>
  );
};

const Child2 = (props) => {
  const { child2Val, callback } = props;
  const onClick = () => {
    const val = Math.random() * 1000;
    callback && callback(val);
  };
  return (
    <>
      <span>我是子组件 2</span>
      <span>{child2Val}</span>
      <button onClick={onClick}>点我回调父组件改变 Child1 的值</button>
    </>
  );
};

const App = () => {
  const [child1Val, setChild1Val] = useState(0);
  const [child2Val, setChild2Val] = useState(0);
  const child1Callback = (val) => {
    setChild2Val(val);
  };
  const child2Callback = (val) => {
    setChild1Val(val);
  };
  return (
    <>
      <span>我是父组件</span>
      <Child1 child1Val={child1Val} callback={child1Callback} />
```

```
      <Child2 child2Val={child2Val} callback={child2Callback} />
    </>
  );
};
```

上述代码中实现的了 Child1、Child2 两个子组件在组件内部点击按钮修改对方展示值的功能，核心思路整理如下：

- 由父组件管理两个子组件的所有状态；
- 发起跨子组件的状态更改起始于子组件内部，回调给父组件；
- 父组件只提供状态的管理和中转，没有业务逻辑。

该方案就是把通信的数据和方法都定义在父组件中，通过 props 传递到子组件中，兄弟子组件分别调用对应的方法即可。

还有一种方案就是发布订阅模式，如果你掌握了 JavaScript 设计模式，那一定不陌生，这其实并不是 React 内置的解决方案，而是跨框架常用的编程方案。例如在一个业务场景中，有 A 和 B 两个表格，互为兄弟组件（也可以是其他关系），当 A 中某个事件调用完成后需要触发 B 表格的数据更新，使用发布-订阅模式也是个很好的选择。首先创建一个发布订阅类，并且直接挂载到 window 对象中：

```
class SubscriptionPublish {
  private eventMap: Record<string, ((params: any) => any)[]>;
  constructor() {
    this.eventMap = {};
  }
  /**
   * 订阅函数
   * @param key 订阅事件 Key 值
   * @param handler 订阅事件
   */
  on(key: string, handler: (params: any) => any) {
    if (!this.eventMap[key]) {
      this.eventMap[key] = [];
    }
    this.eventMap[key].push(handler);
  }
  /**
   * 发布函数
   * @param key 订阅事件 Key 值
   * @param params 要发布到订阅事件中的参数
   */
  emit(key: string, params?: any) {
```

```
      if (this.eventMap[key]) {
        this.eventMap[key].forEach((handler) => {
          handler(params);
        });
      }
    }
    /**
     * 销毁函数
     * @param key
     * @param handler
     */
    remove(key: string, handler: (params: any) => any) {
      if (this.eventMap[key]) {
        const res = this.eventMap[key].indexOf(handler);
        res !== -1 && this.eventMap[key].splice(res, 1);
      }
    }
}
// 创建一个实例挂载到 window 中
window.subscriptionPublish = new SubscriptionPublish();
```

在 SubscriptionPublish 类中实现了三个方法：

● on: 订阅函数，用于监听某个事件，对应依赖于表格 A 的某个事件触发执行更新的表格 B，直接在表格 B 中执行订阅逻辑；

● emit: 发布函数，用于发布某个事件，对应表格 A 更新完毕后需要让表格 B 执行更新，在表格 A 主动端进行发布逻辑；

● remove: 销毁函数，当业务逻辑结束，如页面销毁后，将发布的事件销毁，可以放在 useEffect 的 return 中。

使用方式也很简单，表格 B 在页面挂载时订阅监听事件，当表格 A 事件完成执行发布时，表格 B 更新，代码块如下：

```
const BTable = () => {
  // 刷新列表方法
  const loadTable = () => {
    table.reload();
  };
  useEffect(() => {
    // 订阅刷新方法
    window.subscriptionPublish.on('loadTable', loadTable);
    return () => {
      // 组件销毁时，销毁订阅函数
```

```
      defaultEvent.remove('loadTable', loadTable);
    };
  }, []);
  return <>{/* ... */}</>;
};
```

表格 A 在业务逻辑结束后进行事件的发布即可更新表格 B，代码块如下：

```
const ATable = () => {
  // 触发 B 表格更新
  const loadTable = () => {
    window.subscriptionPublish.emit('loadTable');
  };
  return (
    <>
      <span onClick={loadTable}>更新 B 表格</span>
    </>
  )
};
```

这样的业务场景其实有很多，因此发布-订阅模式是实战中比较常见的方案。该方案并不局限于 React，在其他框架中也可使用。

2.2.7 跨组件分层通信

在关系复杂的组件中，共享数据通信的方式比较常用的有两种方案：createContext+useContext 跨组件数据透传，或全局状态管理。

设想有如图 2.7 所示的一个页面结构。

面对这样错综复杂的组件结构，如果在 App 中获取到数据，App 下级的所有子组件都需要消费数据，该如何处理？若用发布-订阅模式，需要为每个组件都注册监听；若用 props 就更加复杂了，并且也不利于维护。因此，React 针对这种场景提供了内置的 createContext+useContext 跨组件数据透传方案，接下来详细介绍一下其应用及其原理。

在 React 中提供了一种数据管理机制：React.context（简称 Context）。Context 提供了一种在组件树中共享数据的方式，避免了通过 props 逐层传递数据的繁琐过程。它特别适用于那些需要在多个层级之间传递全局数据（如主题、用户信息等）的场景。下面以 Hook 函数组

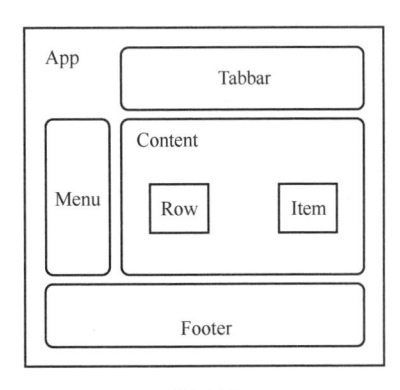

图 2.7

件为例，展开介绍 Context 的使用。

首先需要创建一个 Context 对象，该对象可在类似图 2.7 中的 App 组件中去定义声明：

```
export const Context = React.createContext(defaultValue);
```

当 React 渲染了这个 Context 对象的组件，这个组件就会在组件树中的 Context.Provider 中读取到 defaultValue 的值。

接下来用 Context.Provider 包裹 App 组件，value 属性就是透传给下游所有子组件值的来源，当 value 中的值在 App 组件中发生状态变化时，下游所有消费组件都会重新渲染：

```
<Context.Provider value={/* 某个值，一般会传递对象 */}>
  <div id="app">我是 App 组件</div>
</Context.Provider>
```

那么，下游的子组件该如何消费数据呢？对于函数组件，可通过 useContext API 得到该值。

```
import { Context } from './App.tsx';
const value = React.useContext(Context);
```

useContext 函数接收一个 Context 对象（在 App 组件中 React.createContext 的返回值），返回 Context 的当前值。当组件上游的 Context.Provider 更新时，当前组件会触发重渲染，并读取最新传递给 Context Provider 的 value。

最后通过一个简单示例来总结上述 Context 的使用：

```
const Context = React.createContext(null);
const Child = () => {
  const value = React.useContext(Context);
  return <div>theme: {value.theme}</div>;
};
const App = () => {
  const [theme, setTheme] = React.useState('light');
  const toggleTheme = () => {
    if (theme === 'light') {
      return setTheme('dark');
    }
    setTheme('light');
  };
  return (
    <Context.Provider value={{ theme }}>
      <div onClick={toggleTheme}>切换主题</div>
```

```
      <Child />
    </Context.Provider>
  );
};
ReactDOM.render(<App />, document.getElementById('root'));
```

示例中，在 App 组件内使用 Provider 将应用主题通过 value 向子级传递，Child 组件通过 useContext 读取 value，从而消费传递过来的 theme。

从使用层面分析，可见 Context 的实现由三部分组成：

- 创建 Context：React.createContext()方法；
- Provider 组件：<Context.Provider value={value}>；
- 消费 value：React.useContext(Context)。

原理分析脱离不了源码，下面我们挑选出核心代码简单看一下实现方式。先从 createContext 函数开始，其源码在 react/src/ReactContext.js 中。代码如下：

```
const REACT_PROVIDER_TYPE = Symbol.for('react.provider');
const REACT_CONTEXT_TYPE = Symbol.for('react.context');
export function createContext<T>(defaultValue: T): ReactContext<T> {
  const context: ReactContext<T> = {
    $$typeof: REACT_CONTEXT_TYPE,
    _calculateChangedBits: calculateChangedBits,
    // 并发渲染器方案，分为主渲染器和辅助渲染器
    _currentValue: defaultValue,
    _currentValue2: defaultValue,
    _threadCount: 0, // 跟踪此上下文当前有多少个并发渲染器
    Provider: (null: any),
    Consumer: (null: any),
  };
  context.Provider = {
    $$typeof: REACT_PROVIDER_TYPE,
    _context: context,
  };
  context.Consumer = context;
  return context;
};
```

这段代码逻辑比较简单，但无法看实现数据传输的逻辑，故暂时只需关注该函数的入参和出参数据结构即可。React.createContext 创建出来的节点如图 2.8 所示。

可以看到$$typeof 类型为 react.context 类型，再通过 React.createElement 把 Provider 节点打印出来，如图 2.9 所示。

可见，对象中的 props 保存了 Context 向下传递的 value，而对象的 type 保存

的则是 context.Provider。至此，数据源的奥秘解开了。

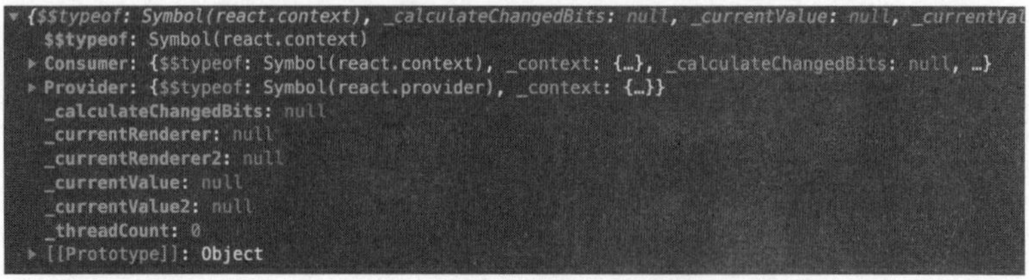

图 2.8

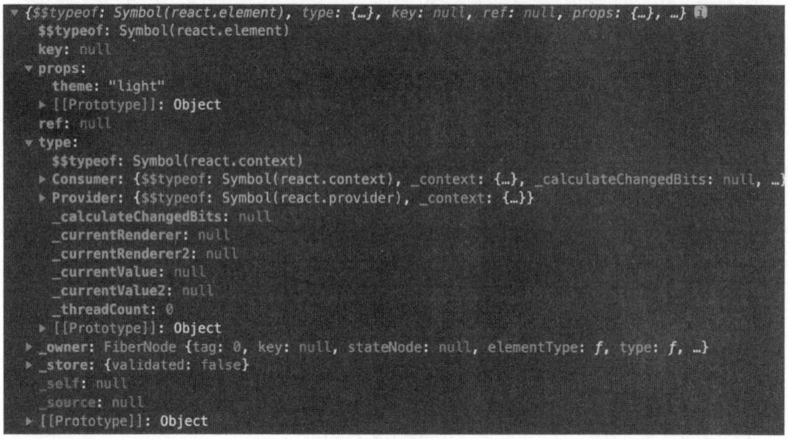

图 2.9

```
context.Provider = {
  $$typeof: REACT_PROVIDER_TYPE,
  _context: context,
};
```

有了对象描述结构，接下来进入渲染流程。React 通过 createElement 创建完所有节点后就是调度过程，将节点树转换为 Fiber 树，在 Reconciler/beginWork 阶段为其创建 Fiber 节点。

useContext 接收 Context 对象作为参数，从 context._currentValue 中读取消费的 value 值。Context 内容除了保存在 value 中以外，还会保存在组件的 Fiber.dependencies 上。其目的是当 Provider value 发生变化时，可以查找到消费 Context 的子组件并执行组件的重渲染。useContext 源码如下：

```
function useContext(context) {
    // 将 context 记录在当前 Fiber.dependencies 节点上
    // 在 Provider 检测到 value 更新后，会查找消费组件标记更新
    const contextItem = {
```

```
    context: context,
    next: null, // 一个组件可能注册多个不同的 context
  };
  if (lastContextDependency === null) {
    lastContextDependency = contextItem;
    currentlyRenderingFiber.dependencies = {
      lanes: NoLanes,
      firstContext: contextItem,
      responders: null
    };
  } else {
    lastContextDependency = lastContextDependency.next = contextItem;
  }
  return context._currentValue;
}
```

看到这里，相信你对于 createContext+useContext 实现跨组件通信的原理有了更深刻的了解。对于另一种方案——全局状态管理，其通常需引用第三方包，如 react-redux、Redux、zustand 等，大体思路是整个应用有一个全局状态库，对该库进行全局数据的读取与更改，这些内容将在第 4 章进行具体讲解。

2.3 组件生命周期

React 中的生命周期是指 React 组件从创建、更新到卸载的整个过程。其可分为以下几个阶段：
- 挂载阶段：组件实例被创建和插入 DOM 树的过程；
- 更新阶段：组件被重新渲染的过程；
- 卸载阶段：组件从 DOM 树中被删除的过程。

由于 React 16.8 版本正式发布了 Hook 机制，并且函数式组件成为了主流开发范式，因此本节会分别讲解类组件和函数组件的生命周期。

2.3.1 类组件生命周期

在类组件中，组件生命周期由一个个独立的生命周期钩子函数组成，总结如下：
- constructor：初始化状态；
- componentWillMount：组件将要挂载时触发的函数；
- render：组件挂载时触发的函数；
- componentDidMount：组件挂载完成时触发的函数；

- componentWillUnmount：组件将要销毁时触发的函数；
- componentWillReceiveProps(nextProps)：父组件中改变 props 传值时触发的函数；
- shouldComponentUpdate(next, nextState)：判断是否要更新组件时触发的函数；
- componentWillUpdate(nextProps, nextState)：将要更新组件时触发的函数；
- componentDidUpdate：组件更新完成时触发的函数；
- getDerivedStateFromProps(nextProps, prevState)：静态方法生命周期钩子函数；
- getSnapShotBeforeUpdate(prevProps, prevState)：保存状态快照。

以下将按如上顺序逐个讲解。

① 在类组件中，React 会在 constructor 中完成数据的初始化，与普通的类一样，它接收两个参数——props 和 context，同时使用 super 声明，就像这样：

```
class App extends React.Component {
  constructor(props) {
    super(props);
    this.state = { name: 'Jack' };
  }
  // ...
}
```

需注意，只要使用了 constructor 就必须写 super，否则会导致类组件中 this 指向错误。

② componentWillMount 函数是挂载到 DOM 之前的钩子函数，使用率较少，更多的使用场景是在服务端渲染，代表组件已经经历了 constructor 初始化数据，但是 DOM 还未渲染出来的时机。

③ render 函数是类组件中唯一必须声明的方法，而其余生命周期钩子函数都是可选的，因为组件渲染必定要走到该生命周期，组件展示的内容也由 render 函数的返回值来决定。render 函数会插入 JSX 生成的 DOM 结构，React 会生成一份虚拟 DOM 树，在每一次组件更新时都会重新进行 diff 算法比较，最终再调用 render 函数重新渲染，若有更新，则有 render。

④ componentDidMount 函数使用率比较高，代表了组件挂载到 DOM 之后的生命周期钩子函数，当组件第一次渲染完成后，DOM 节点生成完毕，所有页面加载的网络请求基本都在这里执行，返回数据后 setState 会重新渲染。

⑤ componentWillUnmount 函数代表组件卸载之前的生命周期钩子函数，经常

会在这里进行一些组件中定时器或全局事件的回收，避免造成内存泄漏，节约性能。

⑥ componentWillReceiveProps 在接收父组件改变后的 props 需要重新渲染子组件时用得比较多，它会接收一个参数，为更新后的 props，可以通过对比 this.props 和 nextProps 来决定是否要重新渲染组件，就像这样：

```
class App extends React.Component {
  componentWillReceiveProps(nextProps) {
    if (nextProps.isOpen !== this.props.isOpen) {
      this.setState({
        isOpen: nextProps.isOpen,
      });
    }
  }
  // ...
}
```

⑦ shouldComponentUpdate 函数更像是一个开关，表示决定是否要更新组件，主要用来优化性能，减少更新，它暴露给了开发者 nextProps、nextState，通过对比即将更新的 state、props 来判断状态是否发生改变，再返回对应的 true 或 false 决定是否更新。应用场景通常为子组件，因为 React 中父组件状态发生改变会导致子组件也一起更新，因此可以在子组件中加入该生命周期钩子函数来进行判断。

⑧ componentWillUpdate 函数是在组件更新即将完成时触发的，当 shouldComponentUpdate 返回 true 时，组件进入 componentWillUpdate 流程，在这里同样可以得到 nextProps 和 nextState。

⑨ componentDidUpdate 函数则是组件更新完成时触发的函数，它与 componentDidMount 的区别是，后者整个生命周期只会执行一次，而前者每当组件中状态发生变化时，都会执行一次，可作为监听器用于监听状态变更。

⑩ getDerivedStateFromProps 函数会在调用 render 方法之前调用，并在初始挂载及后续更新时都被调用，它返回一个对象来更新 state，如果返回的是 null 则不更新任何内容。

⑪ getSnapShotBeforeUpdate 函数是用来代替 componentWillUpdate 函数的，两者的区别在于：在 React 开启异步渲染模式后，在 render 阶段读取到的 DOM 元素状态不总是与 commit 阶段相同，这就导致了在 componentDidUpdate 中使用 componentWillUpdate 中读取到的 DOM 元素状态是不安全的；getSnapShotBeforeUpdate 会在最终的 render 之前调用，读取到的 DOM 元素状态可以保证与 componentDidUpdate 中保持一致。

整体类组件生命周期流程如图 2.10 所示。

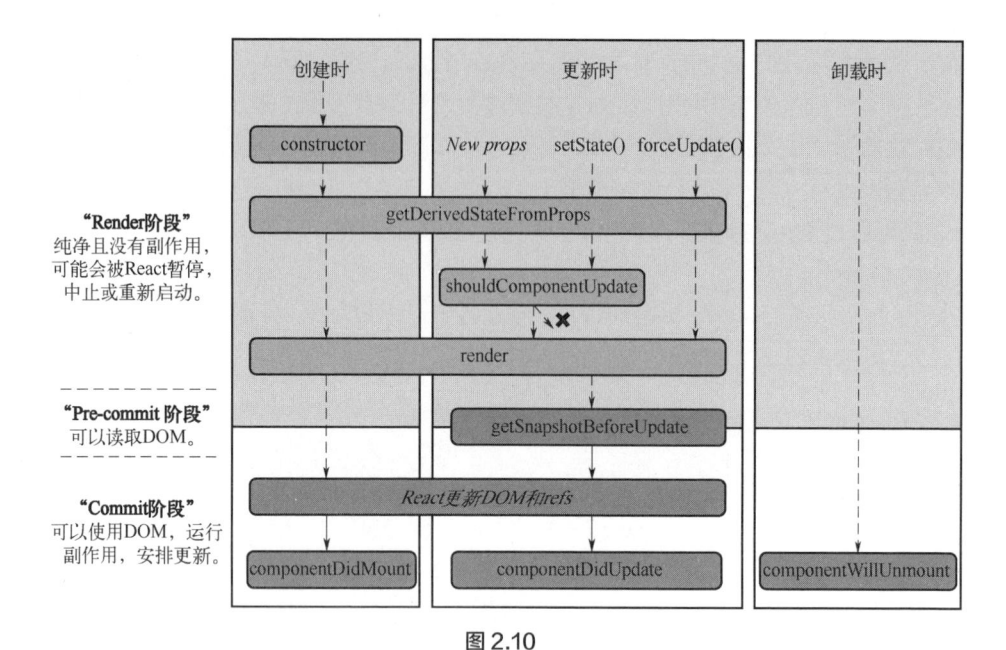

	创建时	更新时	卸载时
	↓	New props setState() forceUpdate()	
	constructor	↓ ↓ ↓	
"Render阶段" 纯净且没有副作用，可能会被React暂停，中止或重新启动。	↓	getDerivedStateFromProps	
	↓	shouldComponentUpdate ✗	
	↓	render	
"Pre-commit阶段" 可以读取DOM。	↓	getSnapshotBeforeUpdate	
"Commit阶段" 可以使用DOM，运行副作用，安排更新。	↓	React更新DOM和refs	
	componentDidMount	componentDidUpdate	componentWillUnmount

图 2.10

讲完了常用的生命周期钩子函数后，一起来看两个实例吧。首先是挂载阶段，创建一个子组件，在父组件中引入函数渲染该组件：

```
class Child extends React.Component {
  // 初始化
  constructor(props) {
    console.log('01 构造函数');
    super(props);
    this.state = {};
  }
  // 组件将要挂载时触发的生命周期函数
  componentWillMount() { console.log('02 组件将要挂载');}
  // 组件挂载完成时触发的生命周期函数
  componentDidMount() { console.log('04 组件挂载完成');}
  // 组件挂载时触发的生命周期函数
  render() {
    console.log('03 数据渲染 render');
    return <div>Child 组件</div>;
  }
}
```

控制台打印结果顺序为：01 构造函数 → 02 组件将要挂载 → 03 数据渲染 render → 04 组件挂载完成。

然后是更新阶段，当组件挂载完成后，组件的状态发生变化，首先

shouldComponentUpdate 确认是否要更新数据，当函数返回 true 才会进行更新。当确认更新数据后，componentWillUpdate 执行，更新完毕后执行 render 函数重新渲染，最后 componentDidUpdate 执行，数据更新完成，完成一次数据更新重渲染。代码如下：

```
class Child extends React.Component {
  constructor(props) {
    super(props);
    this.state = {
      num: 1,
    };
  }
  //是否要更新数据，如果返回 true 才会更新数据
  shouldComponentUpdate(nextProps, nextState) {
    console.log('01 是否要更新数据');
    return true; //返回 true,确认更新
  }
  //将要更新数据时触发的生命周期函数
  componentWillUpdate() { console.log('02 组件将要更新');}
  //更新完成时触发的生命周期函数
  componentDidUpdate() { console.log('04 组件更新完成');}
  //更新数据方法
  onChange() {
    this.setState({
      num: this.num + 1,
    });
  }
  render() {
    console.log('03 数据渲染 render');
    return (
      <div>
        {this.state.num}
        <button onClick={() => this.onChange()}>点我 num+1</button>
      </div>
    );
  }
}
```

控制台打印结果顺序为：01 是否要更新数据 → 02 组件将要更新 → 03 数据渲染 render → 04 组件更新完成。

2.3.2 函数组件生命周期

在 React 16.8 之前，函数组件由于没有生命周期的能力，因此被用来作为展示组件。而 React 推出 React Hooks 之后，赋予了函数组件生命周期的能力，或者更精确地说，其通过 useEffect 使得函数组件能够模拟类组件的生命周期行为。其用法如下。

首先需要引入 useEffect：

```
import { useEffect } from 'react';
```

在函数组件中通过 useEffect 来监听组件挂载和卸载的方式很简单，只需要像这样：

```
useEffect(() => {
  // componentDidMount，挂载，执行副作用
  return () => {
  // componentWillUnmount，将要卸载，清除副作用
  }
}, []); // 传空数组才会引发此监控
```

useEffect 拥有两个参数：第一个参数作为回调函数会在浏览器布局和绘制完成后调用，因此不会阻碍浏览器的渲染进程；第二个参数是个数组，对应触发依赖项。

① 当依赖项有值时，数组中任何值发生改变，都会重新触发执行回调，类似 componentDidUpdate；

② 当依赖项为空数组时，在首次渲染后触发回调，类似 componentDidMount；

③ 当依赖项不存在时，每次渲染都会触发执行一次回调。

useEffect 返回的函数为可选函数，如果返回了，则表示组件即将卸载时的执行动作，类似 componentWillUnmount，可以在这里清除一些组件的副作用，如清除定时器或移除全局事件监听器。

除此以外，我们还可以通过 useState 来为函数组件声明初始化和管理状态，就像这样：

```
const [count, setCount] = React.useState(0);
```

开发者还可以通过 React.memo 包裹一个组件，来模拟 shouldComponentUpdate，通过对比子组件更新前后的 props 来决定是否要更新，第一个参数接收函数组件，第二个参数通过返回 true/false 来决定是否要更新组件，就像这样：

```
// 模拟 shouldComponentUpdate
```

```
const Child = () => {
  // ...
};
React.memo(Child, (prevProps, nextProps) => {
  return nextProps.count !== prevProps.count
});
```

看到这里，读者或许能体会到函数式组件的自由灵活，其更像是原生的 JavaScript 代码，而类组件看起来更加模板化，不利于扩展和维护。函数组件模拟类组件生命周期的对应关系如图 2.11 所示。

类组件	函数组件
constructor	useState
componentDidMount	useEffect第一个参数
componentDidUpdate	useEffect第二个参数监听
componentWillUnmount	useEffect第一个参数返回值
shouldComponentUpdate	memo第二个参数
render	函数返回值
getDerivedStateFromProps	useState返回值的更新方法

图 2.11

2.4 遍历渲染

本节将系统性地介绍在 React 中遍历渲染对象和数组的方式。

2.4.1 遍历渲染对象

对于对象，可以通过 Object.keys()方法得到对象的键值所组成的数组，通过遍历生成的 key 数组来获取具体的 value。同时，将 key 作为 React 遍历元素的 key 属性通常是安全的，因为 JavaScript 中可以确保对象中的键是唯一的，代码块如下：

```
const App = () => {
  const employee = {
    id: 1,
    name: 'Jack',
    age: 25,
  };
  return (
    <div>
```

```
      {Object.keys(employee).map(key => {
        return (
          <div key={key}>
            <h2>
              {key}: {employee[key]}
            </h2>
            <hr />
          </div>
        );
      })}
    </div>
  );
}
```

同样，也可以通过 Object.entries() 方法来生成对象的 key、value，在遍历时通过解构的形式来处理，就像这样：

```
const App = () => {
  const employee = {
    id: 1,
    name: 'Jack',
    age: 25,
  };
  return (
    <div>
      {Object.entries(employee).map(([key, value]) => {
        return (
          <div key={key}>
            <h2>
              {key}: {value}
            </h2>
            <hr />
          </div>
        );
      })}
    </div>
  );
};
```

到了这里你可能会问，难道 Object.values() 不可以吗？当然也可以，前提是需要确保对象中的每个值都是唯一的，并且在未来的迭代中也不会出现问题。有时，甚至还可以使用 index 来作为元素的 key 属性。但是通常还是建议使用遍历体本身更加稳定、独一无二的标识符。

2.4.2 遍历渲染数组

数组的遍历渲染方式有很多，比如普通 for 循环、for-in、for-of、forEach、map、reduce 都可以使用，但是最常用的还是 map，其可读性是最好的。

（1）for 循环遍历渲染列表数据

```
const App = () => {
  // 列表数据
  const listdata = ['苹果', '香蕉', '梨子'];
  // 普通 for 循环
  const oList = [];
  for (let i = 0; i < listdata.length; i++) {
    // 循环列表数据,将每一个数据包裹一层 react 元素,
    // 在 push 到新的 oList 数组中,oList 就是 react 元素列表
    oList.push(<li>{listdata[i]}</li>);
  }
  return <ul> {oList} </ul>;
};
```

（2）for-in 遍历渲染列表数据

```
const App = () => {
  // 列表数据
  const listdata = ['苹果', '香蕉', '梨子'];
  // for-in 循环
  const oList = [];
  for (let index in listdata) {
    // index 为 listdata 列表的索引
    // 通过索引取出数据包裹 react 元素,组成 react 元素列表
    oList.push(<li>{listdata[index]}</li>);
  }
  return <ul> {oList} </ul>;
};
```

（3）for-of 遍历渲染列表数据

```
const App = () => {
  // 列表数据
  const listdata = ['苹果', '香蕉', '梨子'];
  // for-of 循环
  const oList = [];
  for (let value of listdata) {
```

```
  // for-of 直接取出数据,处理为 react 元素.列表
    oList.push(<li>{value}</li>);
  }
  return <ul> {oList} </ul>;
};
```

(4) forEach 遍历渲染列表数据

```
const App = () => {
  // 列表数据
  const listdata = ['苹果', '香蕉', '梨子'];
  // forEach 循环
  const oList = [];
  // forEach 循环将列表数据转为 react 元素列表
  listdata.forEach((value) => {
    oList.push(<li>{value}</li>);
  });
  return <ul> {oList} </ul>;
};
```

(5) map 遍历渲染列表数据

```
const App = () => {
  // 列表数据
  const listdata = ['苹果', '香蕉', '梨子'];
  // map 循环
  // map 返回的就是回调函数 return 内容组成的数组
  const oList = listdata.map((value) => {
    return <li>{value}</li>;
  });

  return <ul> {oList} </ul>;
};
```

(6) reduce 遍历渲染列表数据

```
const App = () => {
  // 列表数据
  const listdata = ['苹果', '香蕉', '梨子'];
  // reduce 循环
  const oList = listdata.reduce((prev, value) => {
    // 没事将数据包裹 react 元素后在 push 到数组中
    prev.push(<li>{value}</li>);
    // 返回数组
    return prev;
```

```
    }, []);
    return <ul> {oList} </ul>;
};
```

2.5 React 事件机制

React 基于浏览器的事件机制自身实现了一套事件机制，包括事件注册、事件合成、事件冒泡和事件派发等，在 React 中这套事件机制被称为合成事件。

合成事件是 React 模拟原生 DOM 事件所有能力的一个事件对象，即浏览器原生事件的跨浏览器包装器根据 W3C 规范来定义合成事件，兼容所有浏览器，拥有与浏览器原生事件相同的接口，例如：

```
const button = <button onClick={handleClick}>按钮</button>
```

如果想要获得原生 DOM 事件，可以通过 e.nativeEvent 属性获取：

```
const handleClick = (e) => console.log(e.nativeEvent);
const button = <button onClick={handleClick}>按钮</button>
```

从上面可以看到 React 事件和原生事件非常相似，但它们也有一定区别：

- 事件名称命名方式不同；
- 事件处理函数书写不同。

```
//原生事件绑定方式
<button onclick="handleClick()">按钮命名</button>;
//React 合成事件绑定方式
const button = <button onClick={handleClick}>按钮命名</button>;
```

虽然 onclick 看似是绑定到 DOM 元素上，但实际上并不会把事件代理函数直接绑定到真实的节点上，而是把所有的事件绑定到结构的最外层，使用一个统一的事件去监听。

这个事件监听器上维持了一个映射来保存所有组件内部的事件监听和处理函数。当组件挂载或卸载时，只是在这个统一的事件监听器上插入或删除一些对象。当事件发生时，首先被这个统一的事件监听器处理，然后在映射里找到真正的事件处理函数并调用。这样做简化了事件处理和回收机制，效率也有了很大提升。

关于 React 合成事件和原生事件的执行顺序，可以参考下面一个例子：

```
import React from 'react';
class App extends React.Component {
  constructor(props) {
      super(props);
```

```
    this.parentRef = React.createRef();
    this.childRef = React.createRef();
  }
  componentDidMount() {
    console.log('React componentDidMount! ');
    this.parentRef.current?.addEventListener('click', () => {
      console.log('原生事件：父元素 DOM 事件监听！');
    });
    this.childRef.current?.addEventListener('click', () => {
      console.log('原生事件：子元素 DOM 事件监听！');
    });
    document.addEventListener('click', (e) => {
      console.log('原生事件：document DOM 事件监听！');
    });
  }
  parentClickFun = () => {
    console.log('React 事件：父元素事件监听！');
  };
  childClickFun = () => {
    console.log('React 事件：子元素事件监听！');
  };
  render() {
    return (
      <div ref={this.parentRef} onClick={this.parentClickFun}>
        <div ref={this.childRef} onClick={this.childClickFun}>
          分析事件执行顺序
        </div>
      </div>
    );
  }
}
export default App;
```

对应的输出顺序如图 2.12 所示。

```
原生事件：子元素 DOM 事件监听！
原生事件：父元素 DOM 事件监听！
React 事件：子元素事件监听！
React 事件：父元素事件监听！
原生事件：document DOM 事件监听！
```

图 2.12

因此可以得出以下结论：React 所有事件都会被挂载到 document 对象上；当对象 DOM 元素触发事件时，会冒泡到 document 对象后，再处理 React 事件；所以会先执行原生事件，然后处理 React 事件；最后真正执行挂载在 document 上的事件。

所以想要阻止不同时间段的冒泡行为，应使用不同的方法：
- 阻止合成事件间的冒泡，用 e.stopPropagation()；
- 阻止合成事件与最外层 document 上的事件间的冒泡，用 e.nativeEvent.stopImmediatePropagation()；
- 阻止合成事件与除最外层 document 上的原生事件上的冒泡，通过判断 e.target 来避免：

```
document.body.addEventListener('click', (e) => {
  if (e.target && e.target.matches('div.code')) {
    return;
  }
});
```

React 事件机制总结如下：

① React 上注册的事件最终会绑定在 document 上，而不是 React 对应的 DOM 上（减少内存开销就是因为所有的事件都绑定在 document 上，其他的节点没有绑定事件）；

② React 自身实现了一套事件冒泡机制，这也是 event.stopPropagation()无效的原因；

③ React 通过队列的形式，从触发的组件向父组件回溯，然后调用它们 JSX 中定义的回调函数；

思考题

1. 类组件和函数组件的区别是什么？哪个是主流选择？为什么？
2. state 和 props 有什么区别？应用场景上有什么本质性区别？
3. 兄弟组件通信相较于父子组件通信会额外用到哪一个特殊的 React API？
4. 类组件有哪些生命周期方法？
5. 函数组件有开箱即用定义明确的生命周期方法吗？如果没有，那是怎么模拟组件生命周期的？
6. 如何通过 map()遍历渲染？在遍历过程中如何避免 key 的冲突？
7. React 事件处理与原生 JavaScript 事件处理有何异同?在 React 中如何阻止事件冒泡？

第 3 章
React Router

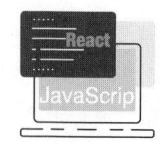

React Router 是一个基于 React 的强大路由库，它可以让开发者向应用中快速添加视图和数据流，同时保持页面与 URL 之间的同步。为了帮助快速了解 React Router 的作用，先列举一个不使用 React Router 的示例：

```
import React, { useState, useEffect } from 'react';
const About = () => {
  return <div>About Component</div>;
};
const Inbox = () => {
  return <div>Inbox Component</div>;
};
const Home = () => {
  return <div>Home Component</div>;
};
const App = () => {
  const [route, setRoute] = useState(window.location.hash.substr(1) || '/');
  useEffect(() => {
    const handleHashChange = () => {
      setRoute(window.location.hash.substr(1));
    };
    window.addEventListener('hashchange', handleHashChange);
    // Clean up the event listener on component unmount
    return () => {
      window.removeEventListener('hashchange', handleHashChange);
    };
  }, []);
  let Child;
  switch (route) {
    case '/about':
      Child = About;
      break;
```

```
    case '/inbox':
      Child = Inbox;
      break;
    default:
      Child = Home;
  }

  return (
    <div>
      <h1>App</h1>
      <ul>
        <li>
          <a href="#/about">About</a>
        </li>
        <li>
          <a href="#/inbox">Inbox</a>
        </li>
      </ul>
      <Child />
    </div>
  );
};
export default App;
```

该 App 组件渲染时监听了哈希路由的变化,当路由变化时,改变组件自身状态,从而渲染不同的<Child />组件。这种方案虽然看着很直接,但是随着每个组件的复杂化,有着更多来自不同 URL 的 UI 组件,为了让 URL 解析变得更智能,需要编写很多与路由相关联的代码,例如某个 URL 指定某个 UI 组件,因此会产生大量的与业务无关的代码。

让我们采用 React Router 来重构这个 demo:

```
import React from 'react';
// 需要导入一些组件...
import { Router, Route, Link } from 'react-router';
// 从应用中删除一堆代码,增加一些<Link>元素
const App = React.createClass({
  render() {
    return (
      <div>
        <h1>App</h1>
        {/*把<a>变成<Link>*/}
        <ul>
```

```
        <li>
          <Link to="/about">About</Link>
        </li>
        <li>
          <Link to="/inbox">Inbox</Link>
        </li>
      </ul>
      {/*接着用`this.props.children`替换`<Child>`
        router 会找到这个 children*/}
      {this.props.children}
    </div>
    );
  },
});

// 最后,我们用一些<Route>来渲染<Router>
// 这些就是路由所提供的我们想要的东西
React.render(
  <Router>
    <Route path="/" component={App}>
      <Route path="about" component={About} />
      <Route path="inbox" component={Inbox} />
    </Route>
  </Router>,
  document.body,
);
```

React Router 知道如何搭建嵌套的 UI,因此我们不需要手动找出需要渲染哪些 <Child /> 组件。对于/about 路径,React 会搭建出<App><About /></App>的映射关系,开发者可以维护一个树级嵌套格式的路由配置表,交给 React Router 去处理:

```
const routes = {
  path: '/',
  component: App,
  childRoutes: [
    { path: 'about', component: About },
    { path: 'inbox', component: Inbox },
  ],
};

React.render(<Router routes={routes} />, document.body);
```

3.1 配置路由

路由配置是一组关系指令，简单理解就是告诉 router 应该如何匹配 URL，即目前是 XXX 路由时应该匹配展示哪个组件，仅此而已。以下通过一个简单例子来解释如何配置路由：

```
import React from 'react';
import { Router, Route, Link } from 'react-router';

const App = React.createClass({
  render() {
    return (
      <div>
        <h1>App</h1>
        <ul>
          <li>
            <Link to="/about">About</Link>
          </li>
          <li>
            <Link to="/inbox">Inbox</Link>
          </li>
        </ul>
        {this.props.children}
      </div>
    );
  },
});

const About = React.createClass({
  render() {
    return <h3>About</h3>;
  },
});

const Inbox = React.createClass({
  render() {
    return (
      <div>
        <h2>Inbox</h2>
        {this.props.children || 'Welcome to your Inbox'}
      </div>
```

```
      );
    },
});

const Message = React.createClass({
  render() {
    return <h3>Message {this.props.params.id}</h3>;
  },
});

React.render(
  <Router>
    <Route path="/" component={App}>
      <Route path="about" component={About} />
      <Route path="inbox" component={Inbox}>
        <Route path="messages/:id" component={Message} />
      </Route>
    </Route>
  </Router>,
  document.body,
);
```

通过上面的配置，这个 React 程序已经知道了如何渲染下面四个 URL 与组件之间的映射关系（表 3.1）：

表 3.1　URL 与组件之间的映射关系

URL	组件
/	App
/about	App → About
/inbox	App → Inbox
/inbox/messages/:id	App → Inbox → Message

如果在迭代系统时，想要将某个页面的路由修改，那么用户通过旧的链接进入系统将会看到一个错误的页面。此时，可以对旧的链接进行重定向，使其映射至新的链接，就像这样：

```
React.render(
  <Router>
    <Route path="/" component={App}>
      <Route path="about" component={About} />
      <Route path="new" component={New} />
      {// 重定向}
```

```
      <Redirect from="old" to="new" />
      <Route path="inbox" component={Inbox}>
        <Route path="messages/:id" component={Message} />
      </Route>
    </Route>
  </Router>,
  document.body,
);
```

这样，当用户访问旧的项目地址时也会自动重定向到新的页面。因为 route 一般被嵌套使用，所以使用 JSX 这种具有简洁嵌套式语法的结构来描述它们的关系非常友好，当然如果不想使用 JSX，也可以直接使用原生 route 数组对象来描述。上面的例子可以被写成下面这样：

```
const routeConfig = [
  { path: '/',
    component: App,
    childRoutes: [
      { path: 'about', component: About },
      { path: 'inbox',
        component: Inbox,
        childRoutes: [
          { path: 'messages/:id', component: Message },
        ]
      }
    ]
  }
];

React.render(<Router routes={routeConfig} />, document.body);
```

3.2 React Router 实现原理

React Router 是一种流行的前端路由解决方案，它允许我们在单页应用程序（SPA）中处理 URL 更改并相应地呈现不同的 React 组件。React Router 基于两个主要的历史模式：HTML5 history 模式（历史路由）和 hash 模式（哈希路由）。

HTML5 history 模式使用 HTML5 history API 来改变 URL，并且在浏览器的地址栏上显示新的 URL，而不需要重新加载页面。要改变 URL，可以使用以下代码：

```
history.pushState(state, title, path);
```

- state：一个代表网址相关的状态对象，在 pushstate 事件被触发时，该对象会传入回调函数，如果不需要此参数，可以设置为 null；
- title：新页面的标题，可填 null；
- path：新的网址，必须与当前页面在同一个域，浏览器的地址栏将显示这个地址。

也可以通过另一种方式直接改变 history 对象记录，history.length 的长度不会发生改变：

```
history.replaceState(state, title, path);
```

当 URL 改变时，浏览器会触发 popstate 事件。我们可以通过添加事件监听器来捕获这些变化：

```
window.addEventListener('popstate', function (e) {
  /* 监听改变 */
});
```

同一个文档的 history 对象发生变化时，就会触发 popstate 事件。采用 history.pushState 或者 history.replaceState 不会触发 popstate 事件，popstate 事件只会在浏览器某些行为下触发，比如点击后退、前进按钮，或者调用 history.back()、history.forward()、history.go() 方法。history.pushState 可以使浏览器地址改变，但是无需刷新页面。

而 hash 模式则通过 window.location.hash 来获取页面的 hash 值，通过 onhashchange 来监听哈希路由的变化：

```
window.addEventListener('hashchange', function (e) {
  /* 监听改变 */
});
```

React Router 的核心依赖于名为 "history" 的独立 JavaScript 库。这个库提供了一种在各种浏览器和环境中管理历史记录的通用 API。具体来说，history 库提供了三种类型的历史记录管理 API：

- browserHistory：基于 HTML5 的 history API，适用于现代浏览器。
- hashHistory：基于 URL 哈希，适用于旧版浏览器。
- memoryHistory：基于内存，适用于非浏览器环境（如 Node.js 环境）或测试场景。

这三种不同环境的 API 有一些共性的操作，将其抽象到了一个公共文件方法——createHistory 中：

```
// 内部的抽象实现
```

```
function createHistory(options={}) {
  ...
  return {
    listenBefore, // 内部的 Hook 机制,可在 location 发生变化前执行某些行为
    listen, // location 发生改变时触发回调
    transitionTo, // 执行 location 的改变
    push, // 改变 location
    replace,
    go,
    goBack,
    goForward,
    createKey, // 创建 location 的 key,用于唯一标示该 location,随机生成
    createPath,
    createHref,
    createLocation, // 创建 location
  }
}
```

每个 API 都会根据需要覆盖其中的一些方法。值得注意的是,这里的 location 与浏览器的原始 location 不同,主要的区别在于多了一个名为 key 的字段,这是 history 内部用于 location 操作的关键。

createLocation()方法返回一个新的 location 对象,包含以下属性:

- pathname: url 的基本路径。
- search: 查询字段。
- hash: url 中的哈希值。
- state: url 对应的 state 字段。
- action: 分为 push、replace 和 pop 三种。
- key: 生成方法为 Math.random().toString(36).substr(2, length)。

现在让我们看看各个 API 的具体实现:

- createBrowserHistory: 利用 HTML5 中的 history API。
- createHashHistory: 通过 hash 存储不同状态下的 history 信息。
- createMemoryHistory: 在内存中进行历史记录的存储。

URL 前进对应的底层接口如下:

- createBrowserHistory: pushState 和 replaceState。
- createHashHistory: location.hash=和 location.replace()。
- createMemoryHistory: 在内存中进行历史记录的存储。

例如,createBrowserHistory 中的前进实现可能如下:

```
// createBrowserHistory(HTML5)中的前进实现
```

```
function finishTransition(location) {
  ...
  const historyState = { key };
  ...
  if (location.action === 'PUSH') {
    window.history.pushState(historyState, null, path);
  } else {
    window.history.replaceState(historyState, null, path)
  }
}
```

createHashHistory 的内部实现可能如下:

```
// createHashHistory 的内部实现
function finishTransition(location) {
  ...
  if (location.action === 'PUSH') {
    window.location.hash = path;
  } else {
    window.location.replace(
    window.location.pathname + window.location.search + '#' + path
    );
  }
}
```

createMemoryHistory 的内部实现可能如下:

```
entries = [];
function finishTransition(location) {
  ...
  switch (location.action) {
    case 'PUSH':
      entries.push(location);
      break;
    case 'REPLACE':
      entries[current] = location;
      break;
  }
}
```

检测 URL 改变的底层接口如下:

- createBrowserHistory: popstate。
- createHashHistory: hashchange。
- createMemoryHistory: 在内存中操作,与浏览器无关,因此不需要触发 UI

层面的事件。

伪代码实现如下:

```
// createBrowserHistory(HTML5)中的后退检测
function startPopStateListener({ transitionTo }) {
  function popStateListener(event) {
    ...
    transitionTo( getCurrentLocation(event.state) );
  }
  addEventListener(window, 'popstate', popStateListener);
  ...
}

// createHashHistory 的后退检测
function startPopStateListener({ transitionTo }) {
  function hashChangeListener(event) {
    ...
    transitionTo( getCurrentLocation(event.state) );
  }
  addEventListener(window, 'hashchange', hashChangeListener);
  ...
}

// createMemoryHistory 的内部实现
function go(n) {
  if (n) {
    ...
    current += n;
    const currentLocation = getCurrentLocation();
    // 改变动作为 POP
    history.transitionTo({ ...currentLocation, action: POP });
  }
}
```

对于 state 状态的维护和存储,可以直接存储在 sessionStorage 中:

```
// createBrowserHistory/createHashHistory 中 state 的存储
function saveState(key, state) {
  ...
  window.sessionStorage.setItem(createKey(key), JSON.stringify(state));
}

// createMemoryHistory 仅仅在内存中
```

```
const storage = createStateStorage(entries);
function saveState(key, state) {
  storage[key] = state
}
```

基于 API 层面来描述具体实现过程（以 BrowserHistory 为例）可以总结如下：
- 点击 Link 标签触发 URL 变化，得到新的 location；
- sessionStorage 保存新的路由 state；
- 执行原生 history.pushState 操作；
- 执行回调，开始进行路由匹配，最后切换到对应新路由的组件，触发重新渲染；
- 实现 UI 更新。

实现的过程如图 3.1 所示。

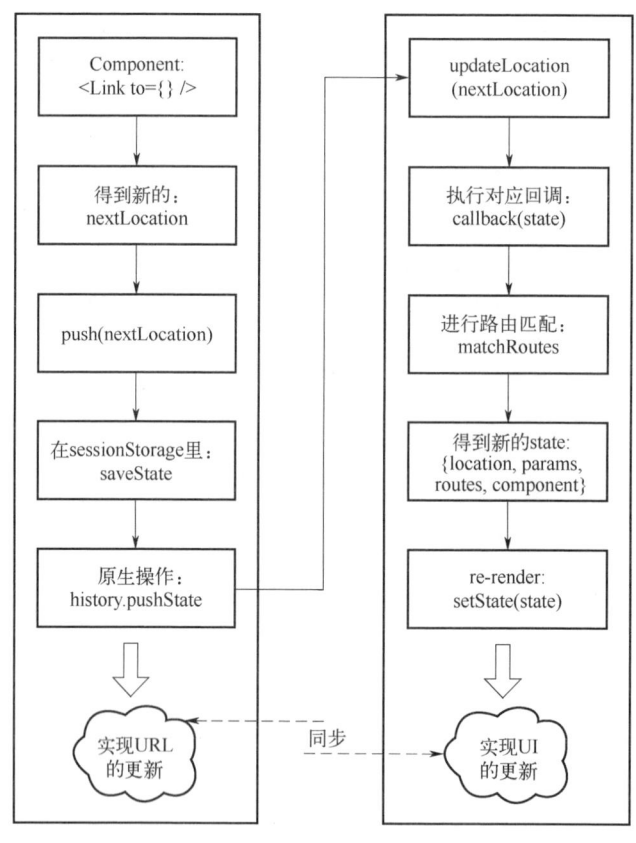

图 3.1

React Router 是一个广泛应用于实际项目中的路由管理库，它与 React 的设计理念高度融合，使用起来自然流畅。它无需手动维护路由状态，代码简洁且易于理解。同时，React Router 提供了多样化的路由配置方式，既可以使用声明式组件的方式，也可以通过配置对象来定义路由规则，极大地提升了开发的灵活性和效率。然而，React Router 的版本升级迭代时 API 变化较大，对旧版本不友好，升级成本较高。

3.3 React Router V6 详解

随着 React Router 官方的更新，从 React Router V5 版本开始已经放弃原有的 react-router 库，目前最新稳定版本（V6）中已经统一命名为 react-router-dom。本节将从 React Router V6 的特性角度来讲解。

首先，通过 create-react-app 工具创建一个 React 项目：

```
create-react-app router-project
```

然后启动全局路由模式，通常有 HashRouter 和 BrowserRouter 两种可选，这里采用 BrowserRouter 的方式，项目 index.js 代码如下：

```
import React from 'react';
import ReactDOM from 'react-dom';
import './index.css';
import App from './App';
import reportWebVitals from './reportWebVitals';
import { BrowserRouter } from 'react-router-dom';

ReactDOM.render(
  <React.StrictMode>
    <BrowserRouter>
      <App />
    </BrowserRouter>
  </React.StrictMode>,
  document.getElementById('root'),
);

reportWebVitals();
```

在项目终端输入命令：npm run start，就可以在浏览器中访问到这个组件了。

React Router V6 在常用路由组件的基础上加入了 Hook 的支持，常用路由组件的能力如表 3.2 所示，常用的 Hook 如表 3.3 所示。

下面的代码创建了两个组件 Home 和 About，然后分别注册了 "/" 和 "/about"

两个路由，在每个页面中还有 Link 组件来进行导航：

表 3.2 常用路由组件

组件名	作用	说明
<Routers>	定义路由配置	代替旧版本<Switch>，所有子路由都用基础 Router children 来表示
<Router>	基础路由	渲染路由组件（可嵌套使用）
<Link>	导航组件	跳转使用
<Outlet>	自适应渲染组件	根据实际路由 URL 自动选择组件

表 3.3 常用路由相关 Hook

Hook	作用	说明
useParams	返回当前参数	根据路径读取参数
useNavigate	返回当前路由	代替原 React Router V5 中的 useHistory
useOutlet	返回根据路由生成的 element	
useLocation	返回当前的 location 对象	
useRoutes	同 Routers 组件一样，只是变成了 Hook 的写法	
useSearchParams	用来匹配 URL 中 "?" 后面的参数	

```
import './App.css';
import { Routes, Route, Link } from 'react-router-dom';

function App() {
  return (
    <div className="App">
      <header className="App-header">
      <Routes>
        <Route path="/" element={<Home />}></Route>
        <Route path="/about" element={<About />}></Route>
      </Routes>
      </header>
    </div>
  );
}
function Home() {
  return (
    <div>
      <main>
        <h2>Welcome to the homepage</h2>
      </main>
      <nav>
```

```
      <Link to="/about">about</Link>
    </nav>
  </div>
 );
}
function About() {
  return (
    <div>
      <main>
        <h2>Welcome to the about page</h2>
      </main>
      <nav>
        <ol>
          <Link to="/">home</Link>
          <Link to="/about">about</Link>
        </ol>
      </nav>
    </div>
  );
}
export default App;
```

运行后的页面如图 3.2 所示。

图 3.2

嵌套路由是 V6 版本对之前版本的一个较大升级，采用嵌套路由会被更智能地识别。假设有如下的路由架构：

```
function App() {
  return (
```

```
    <Routes>
      <Route path="user" element={<Users />}>
        <Route path=":id" element={<UserDetail />} />
        <Route path="create" element={<NewUser />} />
      </Route>
    </Routes>
  );
}
```

当访问路由/user/123 的时候，渲染的组件树会变成这样：

```
<App>
  <Users>
    <UserDetail />
  </Users>
</App>
```

当访问路由/user/create 的时候，组件树将变成这样：

```
<App>
  <Users>
    <NewUser />
  </Users>
</App>
```

如果只是组件内部修改，也可以采用<Outlet />来直接实现，就像这样：

```
function App() {
  return (
    <Routes>
      <Route path="user" element={<Users />}>
        <Route path=":id" element={<UserDetail />} />
        <Route path="create" element={<NewUser />} />
      </Route>
    </Routes>
  );
}
function Users() {
  return (
    <div>
      <h1>Users</h1>
      <Outlet />
    </div>
  );
}
```

在 React Router V6 中，默认路由也变得更加简单，index 属性解决了当嵌套路由有多个子路由但本身无法确认默认渲染哪个子路由的问题，就像这样：

```
function App() {
  return (
    <Routes>
      <Route path="/" element={<Layout />}>
        <Route index element={<About />} />
        <Route path="user" element={<User />} />
        <Route path="about" element={<About />} />
      </Route>
    </Routes>
  );
}
```

这样的配置下，当访问/的时候，<Outlet />会默认展示 About 组件作为默认的子组件。

关于通配符，整个 React Router V6 支持以下几种通配符：

- /groups
- /groups/admin
- /users/:id
- /users/:id/messages
- /files/*
- /files/:id/*

对于*路由，最常见的场景就是针对 NotFound 类路由了，即对于未知路由统一匹配到*来展示 404 页面：

```
function App() {
return (
<Routes>
<Route path="/" element={<Home />} />
<Route path="dashboard" element={<Dashboard />} />
<Route path="*" element={<NotFound />} />
</Routes>
);
}
```

对于 React Router V6 新增的 Hook，可以采用 useParams 和 useSearchparams 来更快捷地获取 URL 参数。假设现有 App 路由：

```
function App() {
```

```
  return (
    <Routes>
      <Route path="user" element={<Users />}>
        <Route path=":id" element={<UserDetail />} />
        <Route path="create" element={<NewUser />} />
      </Route>
    </Routes>
  );
}
```

那么在 UserDetail 内部可以通过 useParams 来获取对应的参数：

```
import { useParams } from 'react-router-dom';

export default function UserDetail() {
  let params = useParams();
  return <h2>User: {params.id}</h2>;
}
```

而 useSearchParams 相对复杂，它返回的是一个当前值和 set 方法，就像这样：

```
const [searchParams, setSearchParams] = useSearchParams();
```

使用时可以通过 searchParams.get("id") 来获取参数，同时在页面内也可以显式地通过 setSearchParams({ id: 2 }) 来改变路由，这样当访问 https://URL/user?id=111 时就可以获取和设置路径。

useNavigate 是替代 React Router V5 中 useHistory 的新 Hook，用法和 useHistory 类似，整体使用起来更轻量，使用方式如下：

```
//js写法
let navigate = useNavigate();
function handleClick() {
  navigate("/home");
}
//组件写法
function App() {
  return <Navigate to="/home" replace state={state} />;
}
//替代原有的 go、goBack 和 goForward
<button onClick={() => navigate(-2)}>
Go 2 pages back
</button>

<button onClick={() => navigate(-1)}>
```

```
  Go back
</button>

<button onClick={() => navigate(1)}>
  Go forward
</button>

<button onClick={() => navigate(2)}>
  Go 2 pages forward
</button>
```

3.4 路由守卫

路由守卫是指在路由跳转前、跳转后做一些动作所触发的钩子函数，在后台管理系统中涉及权限控制相关的逻辑时较常见。在实现路由跳转真实动作前，会先校验该登录用户是否有权限，或者是 token 是否过期，校验结果为有权限、未过期才会通过，反之则执行其他操作，例如返回首页或登录页。那么如何通过 React Router 来实现项目中的路由守卫呢？一共有两种方案：

- 通过公共高阶组件拦截；
- 在项目根目录判断拦截。

对于第一种方案，即封装一个高阶组件，将所有渲染真实页面的路由组件传入该高阶组件，在高阶组件中判断权限逻辑。在 React Router 中可以使用 Route 组件的 render 属性或函数式组件来实现路由守卫。使用 render 属性时，可以传入一个函数，在这个函数中实现路由守卫的具体逻辑，从而根据需要渲染不同的组件或页面。例如检查用户是否登录、根据用户角色判断是否拥有该页面的访问权限等。如果不满足条件，可以返回一个权限提示并执行重定向，满足则进行目标组件或页面的渲染。

根据上述思路，我们可以封装一个这样的 RouteComponent 组件，代码如下：

```
import { Route, Redirect } from 'react-router-dom';

const RouteComponent = (props) => {
  const { component: Component, isAuth, ...rest } = props;
  return (
    <Route
      {...rest}
      render={(props) =>
        isAuth ? (<Component {...props} />) : (
```

```
          <Redirect to={{
            pathname: '/login',
            state: { from: props.location },
          }}
          />
        )
      }
    />
  );
};
```

在上面的示例中，RouteComponent 组件接收三个参数：
- component，代表需要渲染的目标组件；
- isAuth，代表是否有权限访问该页面；
- rest，代表其他参数，传入 React Router 的 Route 组件中，例如路由路径。

对于第二种方案，如果使用函数式组件 Hook 写法，可以把路由守卫的判断逻辑写在项目根文件中（通常为 App.jsx 中）。在 useEffect 中，如果不满足权限条件，则通过 history.push 来手动重定向，代码块如下：

```
import { useEffect } from 'react';
import { useHistory, useRoutes, Router } from 'react-router-dom';
import routes from './routes'

const App = () => {
  const history = useHistory();
  const isAuth = true;

  useEffect(() => {
    if (!isAuth) {
      history.push('/login');
    }
  }, [isAuth, history]);

  return <Router>{useRoutes(routes)}</Router>
};
```

这里使用了 React Router V6 的 useRoutes 钩子快速初始化路由列表。在 useEffect 中，如果用户未登录，就会执行 history.push 方法将页面重定向到登录页面。当然，我们也可以优雅一点，根据实际业务场景封装出一个获取权限状态的 useAuth：

```
import { useState, useEffect } from 'react';
import { getUserAuth } from '@/service';
```

```
const useAuth = () => {
  const [auth, setAuth] = useState({
    isLogin: false,
    superAdmin: false,
    userName: '',
  });

  useEffect(() => {
    getUserAuth();
  }, []);

  const getUserAuth = async () => {
    const res = await getUserAuth();
    if (res?.success && res?.data) {
      setAuth(res.data);
    }
  };

  return auth;
};
```

上述代码中，在 useEffect 阶段调用了 getUserAuth 方法来请求服务器，获取当前用户的权限信息，将信息保存在 useAuth 的状态中，向钩子函数外传递出去。

如此，前述在根文件（App.jsx）中实现路由守卫的案例，就可以改造成这样：

```
import { useEffect } from 'react';
import { useHistory, useRoutes, Router } from 'react-router-dom';
import routes from './routes';
import useAuth from './useAuth';

const App = () => {
  const history = useHistory();
  const { isLogin } = useAuth();

  useEffect(() => {
    if (!isLogin) {
      history.push('/login');
    }
  }, [isLogin, history]);

  return <Router>{useRoutes(routes)}</Router>;
};
```

当然，这个 useAuth 只是简单版本，可以通过具体的业务逻辑来改造，比如某个页面只有超管才能访问、某个页面必须登录才能访问等，把所有路由权限相关的逻辑都集成在 useAuth 中，就像这样：

```js
import { useState, useEffect } from 'react';
import { getUserStatus, getUserAuthCodeList } from '@/service';

const useAuth = () => {
  const [auth, setAuth] = useState({
    isLogin: false,
    superAdmin: false,
    userName: '',
  });
  const [pageAuth, setPageAuth] = useState(false);

  useEffect(() => {
    getUserStatus();
    getUserAuthCodeList();
  }, []);

  /**
   * @description: 获取用户状态
   */
  const getUserStatus = async () => {
    const res = await getUserStatus();
    if (res?.success && res?.data) {
      setAuth(res.data);
    }
  };

  /**
   * @description: 获取用户的页面映射权限表，通过当前页面来判断是否有单页面权限
   */
  const getUserAuthCodeList = async () => {
    const res = await getUserAuthCodeList();
    if (res?.success && res?.data) {
      const pathname = location.pathname;
      if (res.data[pathname]) {
        setPageAuth(true);
      }
    }
  };
```

```
  return {
    auth,
    pageAuth,
  };
};
```

如此一来，useAuth 就复杂了起来，需要同时满足用户已登录并且有该页面的权限才能访问，App.jsx 页面部分逻辑就变成了这样：

```
import { useEffect } from 'react';
import { useHistory, useRoutes, Router } from 'react-router-dom';
import { message } from 'antd';
import routes from './routes';
import useAuth from './useAuth';

const App = () => {
  const history = useHistory();
  const { auth, pageAuth } = useAuth();

  useEffect(() => {
    if (!auth.isLogin) {
      history.push('/login');
    }
    if (!pageAuth) {
      history.replace('/');
      message.error('当前页面没有权限');
    }
  }, [auth.isLogin, history]);

  return <Router>{useRoutes(routes)}</Router>;
};
```

若用户未登录，则跳转到登录页；若当前页面没有权限，则返回到系统首页并给出错误提示。

3.5 哈希路由和历史路由

在 React Router 中，哈希路由（HashRouter）和历史路由（BrowserRouter）是两种不同的路由模式，它们主要的区别在于 URL 的表现形式以及适用场景等。

哈希路由通过 URL 的#符号来进行路由，URL 的格式如 http://example.

com/#/path。由于它是基于哈希的,因此它不会引起浏览器的页面重载,也不需要服务器的支持,通常用于不支持服务器配置的静态页面应用中。

而历史路由是利用 HTML5 的 history API 来实现路由,URL 的格式如 http://example.com/path,其看起来更加美观且符合 SEO 标准。历史路由需要服务器的支持,主要体现在刷新页面或者直接访问路径时需要配置服务器,以确保返回正确的 index.html 文件。

两者各有优缺点,哈希路由的优缺点如下:
- 优点:易于使用,无需后端配置,在页面刷新时不会丢失应用状态。
- 缺点:不如历史路由的 URL 美观,影响 SEO。

历史路由的优缺点如下:
- 优点:更加美观的 URL 格式,友好的 SEO,基于原生 HTML5 实现,兼容性更好。
- 缺点:需要后端支持,有一定成本。

在 React Router 中,实现这两种路由都非常简单,只需要通过引入不同的路由组件包裹在最外层即可,如下:

```
import React from 'react';
import { HashRouter , BrowserRouter, Route, Routes, Link } from 'react-router-dom';

function Home() {
  return <h2>Home</h2>;
}

function About() {
  return <h2>About</h2>;
}

function App() {
  return (
    // 若需使用历史路由,使用<BrowserRouter>即可
    <HashRouter>
      <nav>
        <ul>
          <li><Link to="/">Home</Link></li>
          <li><Link to="/about">About</Link></li>
        </ul>
      </nav>
```

```
      <Routes>
        <Route path="/" element={<Home />} />
        <Route path="/about" element={<About />} />
      </Routes>
    </HashRouter>
  );
}

export default App;
```

思考题

1. 如何使用 React Router 实现路由跳转？

2. React Router 的路由跳转和<a>标签跳转有什么区别？又和 React 有什么关联？

3. React Router 是如何处理 URL 变化的？

4. React Router V6 相较于 V5 有哪些改进？

5. 什么是路由守卫？它的作用是什么？如何基于 React Router 设计一个权限控制系统？

6. 哈希路由和历史路由的区别是什么？如果是 SPA 单页面应用，优先推荐使用哪个路由？为什么？

第 4 章

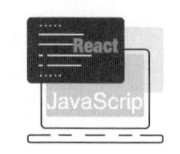

React Hooks 深入浅出

在 React 16.8 版本之前，函数组件通常被称为"无状态组件"，因为它们无法使用 React 的核心特性（如状态管理、生命周期方法等）。而 React 16.8 版本引入了 React Hooks，这一功能使得函数组件能够使用 React 的核心特性，从而大大简化了组件的编写方式，并提升了代码的可读性和可维护性。

但是，使用 Hook 是需要遵循官方的一些规则的：

- 需要避免在循环、条件判断、嵌套函数中调用 Hook；
- 无论发生什么情况，需要保证组件中 Hook 的调用顺序都是稳定唯一的；
- 只有函数定义组件和自定义 Hook 可以调用 React Hooks，避免在类组件或者普通函数中调用；
- 不能在 useEffect 中使用 useState；
- 不需要强制改造类组件，类组件和函数组件两种方式可以并存。

在 React 中，比较重要且常用的 Hook 如下（第 3 章已介绍部分 Hook 的使用方式及原理）：

- useState：用于定义组件的 state 状态，对标到类组件中的 this.state 功能；
- useEffect：通过依赖触发的 Hook，常用于模拟类组件中的 componentDidMount、componentDidUpdate 和 componentWillUnmount 方法；
- useContext：获取来自上游组件的 contenxt 对象，数据透传时会用到；
- useReducer：类似于 Redux 思想的实现，但是并不足以代替 Redux，可以理解为一个组件内部的 Redux，并不是数据持久化存储，会随着组件被销毁而销毁，属于组件内部，配合 useContext 的全局性，可以实现一个轻量局部级别的 Redux；
- useCallback：缓存回调函数，避免传入的回调函数每次都是新的函数，会判断监听依赖项是否发生变化从而重新生成新的函数实例，主要用于性能优化；
- useMemo：与 useCallback 类似，唯一区别是 useCallback 缓存的是函数，而 useMemo 缓存的是具体的值，避免组件每次渲染都重新计算值；
- useRef：获取组件真实节点或者作为组件内部的唯一变量，用法灵活，也可以执行 ref 转发的操作；

● useLayoutEffect：DOM 更新同步 Hook，用法与 useEffect 类似，区别在于执行时间点不同，useEffect 属于异步执行，并不会等待 DOM 真正渲染后执行，而 useLayoutEffect 则会在真正渲染后才触发，可以获取更新后的 state；

● 自定义 Hook：基于 React 内置的一系列 Hook，可以编写适用于业务场景的自定义 Hook。

让我们通过一个简单的按钮组件来感受一下其与 Hook 的区别。先看类组件的实现：

```
import React, { Component } from "react";

export default class Button extends Component {
  constructor() {
    super();
    this.state = { buttonText: "Click me, please" };
    this.handleClick = this.handleClick.bind(this);
  }
  handleClick() {
    this.setState(() => {
      return { buttonText: "Thanks, been clicked!" };
    });
  }
  render() {
    const { buttonText } = this.state;
    return <button onClick={this.handleClick}>{buttonText}</button>;
  }
}
```

这个类组件仅仅是一个按钮，但它的代码已经很"重"了，真实的 React App 由多个类组件按照层级，一层一层构成，复杂度将成倍增长。Redux 的作者 Dan Abramov 曾总结了类组件的几个缺点：

● 大型组件很难拆分和重构，也很难测试；

● 业务逻辑分散在组件的各个方法中，导致出现重复逻辑和关联逻辑；

● 类组件引入了复杂的编程模式，比如 render、props 和高阶组件，增加了开发者的心智负担和上手成本。

将按钮组件采用函数式组件+Hook 的形式重构，代码如下：

```
import React, { useState } from 'react';

export default function Button() {
```

```
  const [buttonText, setButtonText] = useState('Click me, please');
  function handleClick() {
    return setButtonText('Thanks, been clicked!');
  }

  return <button onClick={handleClick}>{buttonText}</button>;
}
```

采用函数式组件+Hook 重构的代码更加简洁、一致，并且符合函数组件的设计理念。

为了更为具体地对比，接下来将分别使用类组件以及函数组件+Hook 的形式实现一个简易的计数器：

```
class Example extends React.Component {
  constructor(props) {
    super(props);
    this.state = {
      count: 0
    };
  }

  render() {
    return (
      <div>
        <p>You clicked {this.state.count} times</p>
        <button
          onClick={() => {
            this.setState({ count: this.state.count + 1 });
          }}
        >
          Click me
        </button>
      </div>
    );
  }
};
```

类组件需要自己声明状态，编写操作状态的方法，而且还需要维护状态的生命周期，显得特别麻烦。如果使用 React Hooks 提供的 useState 来处理状态，代码将会简洁许多，重构后的代码如下：

```
function Example() {
  const [count, setCount] = useState(0);
```

```
    return (
      <div>
        <p>You clicked {count} times</p>
        <button onClick={() => setCount(count + 1)}>
          Click me
        </button>
      </div>
    );
};
```

可以看到，Example 从一个类组件变成了一个函数组件，该函数组件拥有了自己的状态，并且不需要调用 setState() 方法也可以更新自己的状态。

4.1 useState

在了解了 useState 的用法后，为了更好地理解其原理，我们从 React 源码视角看一下 useState 是如何实现的，以及在初始化 state、修改 state、更新 state 时 React 具体做了些什么。

首先找到入口 ReactHooks.js 并从中找到 useState 方法，代码如下：

```
export function useState<S>(initialState: (() => S) | S) {
  const dispatcher = resolveDispatcher();
  return dispatcher.useState(initialState);
}
```

查看 resolveDispatcher 函数的实现，会看到代码块 const dispatcher = ReactCurrentDispatcher.current，继续找到 ReactCurrentDispatcher，源码如下：

```
const ReactCurrentDispatcher = {
  /**
   * @internal
   * @type {ReactComponent}
   */
  current: (null: null | Dispatcher)
};
```

可以看到此时默认的 ReactCurrentDispatcher 是空的，但其会在初始化 state 后进行赋值。再看设置 ReactCurrentDispatcher 的地方，在 ReactFiberScheduler.js 中，源码如下：

```
function renderRoot(root: FiberRoot, isYieldy: boolean): void {
```

第 4 章 React Hooks 深入浅出   101

```
//...
isWorking = true;
// 设置 ReactCurrentDispatcher，包含所有 Hook 的实现
const previousDispatcher = ReactCurrentDispatcher.current;
ReactCurrentDispatcher.current = ContextOnlyDispatcher;
//...
}
```

可以看到 Hook 全程深入参与了整个生命周期。但为什么不在一开始就赋值，而是要在后面赋值呢？因为 Hook 只能在函数式组件中使用，需要通过多次检测判断是否可以使用，而 ReactCurrentDispatcher.current 在执行的时候包含了所有内置 Hook 的实现。

useState 一共分为三个阶段：

- 初始化阶段（mount 阶段），执行 mountState（reactFiberHooks.js），创建一个队列；
- 更改 state 的时候（action 阶段），dispatch（派发）一个包含新值的 action 到队列中，如果当前处于空闲阶段，则提前得到下次的更新作为缓存优化的手段；
- 更新阶段（update 阶段），执行 updateState 方法，循环执行所有的 action，得到最新的一批 state，最后保存到 memorizedState 中。

那么在 React 中是如何判断当前 useState 执行处于哪个阶段的呢？在 ReactFiberHooks.js 文件中有一个 renderWithHooks 方法，有这样一段关键部分代码：

```
// 区分是 Mount 还是 Update
ReactCurrentDispatcher.current =
    nextCurrentHook === null
        ? HooksDispatcherOnMount
        : HooksDispatcherOnUpdate;
```

当 useState 执行时，首先会判断 nextCurrentHook 是否为空，如果为空就默认是第一次初始化挂载，否则就是更新 action。再看一下 HooksDispatcherOnMount 和 HooksDispatcherOnUpdate 两个对象，代码如下：

```
const HooksDispatcherOnMount: Dispatcher = {
  readContext,
  useCallback: mountCallback,
  useContext: readContext,
  useEffect: mountEffect,
  useImperativeHandle: mountImperativeHandle,
  useLayoutEffect: mountLayoutEffect,
  useMemo: mountMemo,
```

```
  useReducer: mountReducer,
  useRef: mountRef,
  useState: mountState,
  useDebugValue: mountDebugValue
};

const HooksDispatcherOnUpdate: Dispatcher = {
  readContext,
  useCallback: updateCallback,
  useContext: readContext,
  useEffect: updateEffect,
  useImperativeHandle: updateImperativeHandle,
  useLayoutEffect: updateLayoutEffect,
  useMemo: updateMemo,
  useReducer: updateReducer,
  useRef: updateRef,
  useState: updateState,
  useDebugValue: updateDebugValue
};
```

这两个对象的方法名一模一样，但实现方式却完全不同。这两者的区别在于：

● HooksDispatcherOnMount：所有的 Hook 都是用来进行初始化的，即一边执行，一边将这些 Hook 添加到单向链表中；

● HooksDispatcherOnUpdate：顺着 Hook 单向链表，按顺序执行，由于 Hook 名称一致，只是执行目的性不一致，所以命名保持统一。

接下来先看 useState 初始化阶段：

```
function mountState<S>(
initialState: (() => S) | S
): [S, Dispatch<BasicStateAction<S>>] {
  // 访问 Hook 链表的下一个节点，获取新的 Hook 对象
  const hook = mountWorkInProgressHook();
  // 获取初始化 state
  // useState 有两种传初始值的方式：直接传值；传入函数，最后返回值
  // 判断如果是函数的话，会自动执行得到返回值
  if (typeof initialState === "function") {
    initialState = initialState();
  }
  // 存入 memoizedState，以在 update 时能得到更新后的值
  hook.memoizedState = hook.baseState = initialState;
  // 初始化队列
  const queue = (hook.queue = {
```

```
    last: null,
    dispatch: null,
    lastRenderedReducer: basicStateReducer,
    lastRenderedState: (initialState: any)
  });
  const dispatch: Dispatch<
    BasicStateAction<S>
  > = (queue.dispatch = (dispatchAction.bind(
    null,
    ((currentlyRenderingFiber: any): Fiber),
    queue
  ): any));
  return [hook.memoizedState, dispatch];
}
```

上例代码依次做了三件事：首先接收一个初始 state 值，创建 Hook 对象，初始值如果是函数，则将函数返回值作为初始值，如果是其他类型，则直接引用，同时代入给 Hook 实例的 memoizedState 和 baseState 属性；然后建立一个 update（更新）队列和 dispatch 触发变更器，将初始值 state 和 dispatch 也就是变量和更改变量的方法最后抛出。

再来看看 dispatch 传入了什么。代码如下：

```
const update: Update<S, A> = {
  expirationTime,
  action,
  eagerReducer: null,
  eagerState: null,
  next: null
};
  // 获取队列中的最后一个更新对象
  const last = queue.last;
  if (last === null) {
    // 如队列为空，将 update 的 next 指向自己，形成一个循环链表
    update.next = update;
  } else {
    // 如果队列不为空，获取队列中的第一个更新对象
    const first = last.next;
    if (first !== null) {
      // 将当前 update 的 next 指向第一个更新对象
      update.next = first;
    }
  // 将当前 update 加入队列
```

```
  last.next = update;
}
```

更改 state 的值其实就是执行了 dispatch（派发），创建了一个 update（更新），将创建的更新放入队列最后，last 会指向 first，会形成环形方便查找，dispatch（派发）一个包含新值的 action 到队列。

那么更新阶段呢？由于在更新阶段最后执行了 scheduleWork，重新经历了一遍调度流程，所以又重新执行 useState 的过程，根据前面的判断进入了更新阶段。再执行 updateState 方法，循环执行所有 update 上的 action，得到最新的 state，然后保存到 memorizedState，并将其返回。

useState 实际上是执行了 updateState，对应挂载阶段的 memoizedState 方法，updateState 方法代码如下：

```
function updateState<S>(
  initialState: (() => S) | S
): [S, Dispatch<BasicStateAction<S>>] {
  return updateReducer(basicStateReducer, (initialState: any));
}
```

updateState 方法其实就是返回了 updateReducer 方法。在 updateReducer 中，其相关代码如下：

```
const hook = updateWorkInProgressHook();
const queue = hook.queue;
```

首先获取了更新队列：

```
renderPhaseUpdates.delete(queue);
let newState = hook.memoizedState;
let update = firstRenderPhaseUpdate;
do {
  const action = update.action;
  // 执行 reducer，得到新的 state，新的 state 再参与计算
  newState = reducer(newState, action);
  // 继续
  update = update.next;
} while (update !== null);
if (!is(newState, hook.memoizedState)) {
  markWorkInProgressReceivedUpdate();
}
// 更新 memoizedState 的值
  hook.memoizedState = newState;
  if (hook.baseUpdate === queue.last) {
```

```
  hook.baseState = newState;
}
queue.lastRenderedState = newState;
// 返回新值
return [newState, dispatch];
```

继续往下，判断 numberOfReRenders > 0 时开始渲染，获取 memoizedState 中保存的 state，进行一个 do while 循环，将所有 action 处理完，循环中执行 reducer 以得到新的 state，循环结束后得到最新的 state，并更新 memoizedState 的值。至此，更新结束。

整体 useState 内部调用更新图如图 4.1 所示。

```
                ┌── mount阶段 ────── 执行mountState(reactFiberhooks.js)，创建一个队列
                │
                │                    dispatch一个包含新值的action到队列
useState ───────┼── 更新state阶段 ──
                │                    如果当前处于空闲状态，则提前得到下次的更新作为缓存优化手段
                │
                └── update阶段 ───── 执行updateState方法，循环执行所有action，得到最新的state，
                                    然后保存到memoizedState
```

图 4.1

4.2 useState 是同步的还是异步的？

useState 是 setState 在函数式组件中实现的 Hook，本质上其实还是执行 setState，因此可先了解 setState 是同步的还是异步的。先看一段代码：

```
constructor(props) {
  super(props);
  this.state = {
    data: 'data'
  }
}

componentDidMount() {
  this.setState({
    data: 'did mount state'
  })

  console.log("did mount state ", this.state.data);
  // did mount state data
```

```
  setTimeout(() => {
    this.setState({
      data: 'setTimeout'
    })

    console.log("setTimeout ", this.state.data);
  });
};
```

这段代码的第一个 console.log 会输出 data,而第二个 console.log 会输出 setTimeout。也就是第一次 setState 的时候,是异步更新的,而第二次 setState 的时候,它又变成了同步更新,是不是有点晕呢? 去源码里看一下 setState 更新调度的时候到底做了些什么。

setState 被调用后最终会执行到 scheduleUpdateOnFiber 函数中,这个函数做了些什么呢?

```
if (executionContext === NoContext) {
  flushSyncCallbackQueue();
}
```

executionContext 代表了 React 目前所处的阶段,而 NoContext 可以理解为是 React 的空闲状态,而 flushSyncCallbackQueue 里面就会去同步调用 this.setState,也就是说同步更新 state。所以,当 executionContext 为 NoContext 的时候,setState 就是同步的。那么,什么情况会改变 executionContext 的值呢?

我们随便找几个地方看看:

```
function batchedEventUpdates$1(fn, a) {
  var prevExecutionContext = executionContext;
  executionContext |= EventContext;
  // ...
}

function batchedUpdates$1(fn, a) {
  var prevExecutionContext = executionContext;
  executionContext |= BatchedContext;
  // ...
}
```

当 React 进入它自己的调度步骤时,会给 executionContext 赋予不同的枚举,表示不同的操作和目前 React 所处的调度状态,而 executionContext 的初始值就是 NoContext,所以只要不进入 React 的调度流程,这个值就是 NoContext,那么 setState

就是同步的。

那在 useState 中呢？自从 React 出了 Hook 之后，函数组件也能拥有自己的状态了，那么如果调用它的第二个参数去 setState 更改状态，和类组件的 this.setState 是一样的效果吗？

没错，因为 useState 的 set 函数最终也会执行到 scheduleUpdateOnFiber，所以在这一点上和 this.setState 是没有区别的，相当于使用了一个通用函数。

但是值得注意的是，当调用 this.setState 的时候，React 会自动帮我们进行一个 state 的合并，而 Hook 则不会，所以我们在使用的时候应着重注意这一点。

举个例子：

```
// 类组件中
state = {
  data: 'data',
  data1: 'data1'
};
this.setState({ data: 'new data' });
console.log(state);
// { data: 'new data',data1: 'data1' }

// 函数组件中
const [state, setState] = useState({ data: 'data', data1: 'data1' });
setState({ data: 'new data' });
console.log(state);
// { data: 'new data' }
```

但是如果你自己去尝试在函数组件中的 setTimeout 中去调用 setState，之后打印 state，会发现并没有改变，这是为什么呢？这不是同步执行的吗？这其实是一个闭包问题，实际上得到的还是上一个 state，那打印出来的值自然也还是上一次的，此时真正的 state 已经改变了。

相信看到这里，对于 "useState 是同步的还是异步的？"，你已经有答案了。只要进入了 React 的调度流程，那就是异步的；只要没有进入 React 的调度流程（executionContext === NoContext），那就是同步的。什么情况下不会进入 React 的调度流程？setTimeout、setInterval，以及直接在 DOM 上绑定原生事件等，这些都不会进入 React 的调度流程，在这种情况下调用 setState，那这次 setState 就是同步的，否则就是异步的。而 setState 同步执行的情况下，DOM 也会被同步执行更新，也就意味着如果多次 setState 会导致多次更新，这也是毫无意义且浪费性能的。

在 setTimeout、原生事件中调用 setState 的操作确实比较少见，还是先看一个案例：

```
const fetch = async () => {
  return new Promise((resolve) => {
    setTimeout(() => {
      resolve('fetch data');
    }, 300);
  })
};

componentDidMount() {
  (async () => {
    const data = await this.fetch();
    this.setState({data});
    console.log("data: ", this.state);
    // data: fetch data
  })()
};
```

这段代码在生命周期 componentDidMount 挂载阶段发送了一个网络请求，收到请求响应结果后再调用 setState，这时候我们用了 async/await 来处理。

此时我们会发现 setState 变成同步的了。componentDidMount 不是 React 的内置 Hook 吗？这难道都不算 React 的调度环境吗？这是因为 componentDidMount 执行完毕后，就已经退出了 React 调度，而请求的代码是异步的，相当于队列中的宏任务还没处理完毕，等结果请求回来以后，setState 才会执行。async 函数中 await 后面的代码其实是异步执行的，这就和在 setTimeout 中执行 setState 是同样的效果，所以 setState 就变成同步的了。

对此，如果变成同步的情况下滥用 setState 会出现什么问题呢？我们来看看在非 React 调度环境下调用 setState 会发生什么：

```
this.state = {
  data: 'init data',
}

componentDidMount() {
  setTimeout(() => {
    this.setState({data: 'data 1'});
    console.log("dom value", document.querySelector('#state').innerHTML);
    this.setState({data: 'data 2'});
    console.log("dom value", document.querySelector('#state').innerHTML);
    this.setState({data: 'data 3'});
    console.log("dom value", document.querySelector('#state').innerHTML);
```

```
    }, 1000)
  }

  render() {
    return (
      <div id="state">
        {this.state.data}
      </div>
    );
  };
```

console 输出的结果如图 4.2 所示。

```
dom value data 1                                              App.js:24
dom value data 2                                              App.js:26
dom value data 3                                              App.js:28
```

图 4.2

可以看到，console 的结果是符合预期的，在 setTimeout 中，属于非 React 调度环境，在 1s 后同步打印了三个最新的结果。

但是界面上从 init data 直接变成了 data 3，这是为什么呢？我们每次都能在 DOM 上得到最新的 state，是因为 React 已经把 state 的修改同步更新了，但是为什么界面上没有显示出来？因为对于浏览器来说，渲染线程和 JS 线程是互斥阻塞的，React 代码运行调度时，浏览器是无法渲染的。所以实际上我们把 DOM 更新了，但是 state 又被修改了，React 只好再进行一次更新，这样反复了三次，最终 React 代码执行完毕后，浏览器才把最终的结果渲染到了页面上，也意味着前两次更新是无意义的。

如果把 setTimeout 去掉，就会发现三次输出都为 init data，因此此时的 setState 就变成了异步的，会把三次更新批量合并到一次去执行，在渲染上也不会出现问题。所以当 setState 变成同步时就要注意，不要写出让 React 多次更新组件的代码，这样是毫无意义的。

React 已经帮助我们做了很多优化措施，但是有时代码不同的实现方式导致了 React 的性能优化失败，相当于我们自己做了反优化，因此深入理解 React 的运行对于日常开发的帮助也是很大的。

4.3　useEffect

React 构建的用户界面由许多组件组成，每个组件处理数据的依赖关系。这些数据在应用程序执行过程中会发生变化，目前通常使用 useState（Hook）来管理这些数据的状态变量。当状态变量发生变化时，组件函数会重新执行并做出相应反应，更新真实 DOM 并重新渲染组件。

当组件中进行 http 请求接口调用时，如果定义了一个函数，那么它会在任何状态变量发生变化时执行 http 调用，导致进入无限循环。为了处理这种情况，React 提供了一个 Hook，即 useEffect。

在深入学习和理解 useEffect 之前，应该先了解一个概念——副作用。在 React 中，副作用指的是与组件渲染结果无关的任何操作，例如：

- 发送网络请求；
- 修改 DOM 元素；
- 访问本地存储；
- 订阅或取消订阅事件；
- 改变组件状态外的变量等。

这些操作会影响组件的行为和状态，但是并不会直接影响渲染结果。在 React 中应该把所有副作用都分离出来，以便于更好地控制组件行为和状态，以保持组件的一致性和可维护性。

useEffect 是一个接收两个参数的函数。传递给 useEffect 的第一个参数是一个名为 effect 的函数，第二个参数是可选的，用于存储依赖关系的数组。下面是其基本用法：

```
import React, { useEffect } from "react";
import { render } from "react-dom";

const App = props => {
  useEffect(() => {
    console.log("Effect has been called");
  }); //省略了 useEffect 的第二个参数
  return <h1> Hello world! </h1>;
};

const root = document.getElementById("root");
render(<App />, root);
```

第一个参数称为 effect，是一个函数，在组件挂载时（第一次渲染时）被执行，在后续的更新中是否被执行由第二个参数传递的依赖关系数组组成，而 effect 函数也可以拥有返回值，返回的是一个函数，称为 cleanup。其执行时机在函数组件卸载前或依赖项变化时执行 useEffect 前，通常用于清理组件产生的全局副作用，如定时器、全局事件等。

第二个参数是一个依赖关系数组，用于控制 effect 函数在组件挂载后什么时候执行，比如：

```
function App(){
  const[state, setState] = useState(0);
  useEffect(() => {
    console.log(state);
    // 此处使用了 state 的值，因此必须将其作为依赖项传递
  }, [state])
};
```

当 state 发生变化时，就会重新执行 effect 函数，打印 state 的新值，React 会比较依赖关系的当前值和之前渲染的值，如果渲染前后的值不一样，就会调用 effect，这个参数是可选的，如果省略它，effect 将在每次渲染后执行。如果想让 effect 只在第一次渲染时执行，可以选择传递一个空数组，就像这样：

```
useEffect(() => {
  console.log("Effect has been called");
}, []) // 使用空数组作为依赖项，useEffect 仅在组件挂载时调用一次
```

读者可以回顾一下 2.3.2 小节通过 useEffect 模拟类组件生命周期的内容，或许会有更多体会。

4.4 useLayoutEffect

在使用层面，useLayoutEffect 的用法完全等同于 useEffect，都是接收一个函数和一个数组，只有在数组的值改变时才会重新执行 effect。但为什么会存在两个用法一样的 Hook 呢？它们的差异主要有两点：

- useEffect 是异步执行的，而 useLayoutEffect 是同步执行的；
- useEffect 的执行时机是浏览器完成渲染之后，而 useLayoutEffect 的执行时机是浏览器把内容真正渲染到界面之前，和 componentDidMount 等价。

用一个很简单的例子来说明这两个 Hook 的区别：

```
import React, { useEffect, useLayoutEffect, useState } from 'react';
import './App.css';

function App() {
  const [state, setState] = useState("hello world")
```

```
useEffect(() => {
    let i = 0;
    while(i <= 100000000) {
      i++;
    };
    setState("world hello");
}, []);

// useLayoutEffect(() => {
//     let i = 0;
//     while(i <= 100000000) {
//       i++;
//     };
//     setState("world hello");
// }, []);

return (
    <>
        <div>{state}</div>
    </>
);
};
```

它的效果是页面从 hello world 闪烁为 world hello，而换成 useLayoutEffect 之后，闪烁现象消失了。

看到这里相信你应该理解它们的区别了。useEffect 是渲染完之后异步执行的，所以会导致 hello world 先被渲染到屏幕上，再变成 world hello，因此出现闪烁现象。而 useLayoutEffect 是渲染之前同步执行的，所以会等待它执行完再渲染上去，就避免了闪烁现象。也就是说开发者最好把 DOM 的相关操作放到 useLayoutEffect 中，避免导致闪烁。

4.5　useEffect 和 useLayoutEffect 的区别

简单来讲，useEffect 和 useLayoutEffect 的区别主要在于执行时机和对浏览器渲染的影响。为了更好地理解它们的区别，有必要先来了解一下 Effect 数据结构，即副作用 Hook（useEffect 和 useLayoutEffect）的类型。

单个 effect 对象包括以下几个属性：
- create: 传入 useEffect 或 useLayoutEffect 函数的第一个参数，即回调函数；
- destory: 回调函数 return 的函数，在该 effect 销毁时执行，渲染阶段为

undefined；
- deps：依赖项，改变后重新执行副作用；
- next：指向下一个 effect；
- tag：effect 的类型，区分是 useEffect 还是 useLayoutEffect。

结合日常应用，单看这些字段还是很通俗易懂的。这里补充一下 Hook 链表的概念，有如下的例子：

```
const Hello = () => {
  const [ text, setText ] = useState('hello')
  useEffect(() => {
    console.log('effect1')
    return () => {
      console.log('destory1');
    }
  })
  useLayoutEffect(() => {
    console.log('effect2')
    return () => {
      console.log('destory2');
    }
  })
  return <div>effect</div>
};
```

挂载到 Hello 组件对应 fiber 节点上的 memoizedState 如图 4.3 所示。可见，memorizedState 的值会根据不同的 Hook 来决定对应的类型。

此外，打印结果和组件中声明 Hook 的顺序是一样的，不难看出来这是一个链表，这也是 React 要求 Hook 的使用不能放在条件分支语句中的原因。如果第一次 mount 进入的是 A 情况，第二次 updateMount 进入的是 B 情况，就会出现 Hook 链表混乱的情况。

- 当使用 useState 时，memorizedState 对应类型为 string（hello），也可以理解成是 state 初始值的类型；
- 当使用 useEffect 和 useLayoutEffect 时，对应的是 Effect 类型。

Hook 类型如下：

```
export type Hook = {
  memoizedState: any, // Hook 自身维护的状态
  baseQueue: any,
  baseState: any,
  queue: UpdateQueue<any, any> | null, // Hook 自身维护更新队列
```

```
  next: Hook | null, // next 指向下一个 Hook
};
```

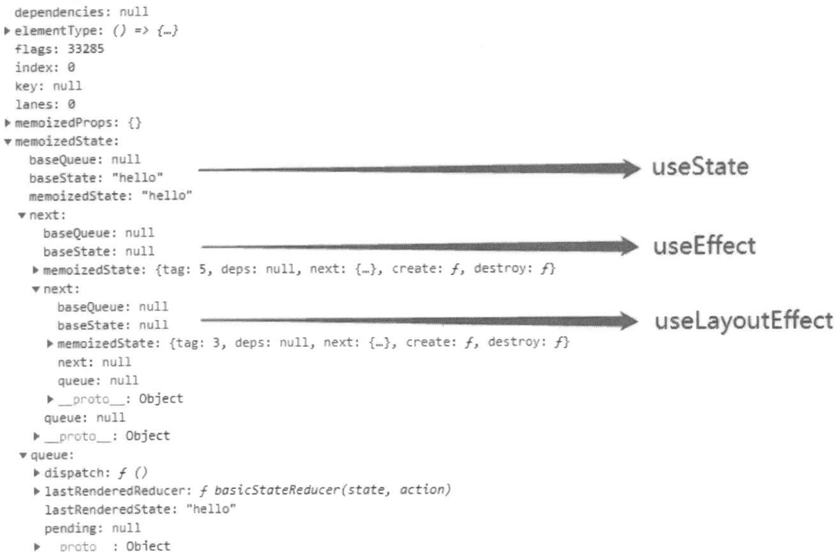

```
dependencies: null
▶ elementType: () => {…}
  flags: 33285
  index: 0
  key: null
  lanes: 0
▶ memoizedProps: {}
▼ memoizedState:
    baseQueue: null
    baseState: "hello"                    ──────────▶ useState
    memoizedState: "hello"
  ▼ next:
    baseQueue: null
    baseState: null                       ──────────▶ useEffect
    ▶ memoizedState: {tag: 5, deps: null, next: {…}, create: f, destroy: f}
    ▼ next:
      baseQueue: null
      baseState: null                     ──────────▶ useLayoutEffect
      ▶ memoizedState: {tag: 3, deps: null, next: {…}, create: f, destroy: f}
      next: null
      queue: null
      ▶ __proto__: Object
    queue: null
    ▶ __proto__: Object
  ▼ queue:
    ▶ dispatch: f ()
    ▶ lastRenderedReducer: f basicStateReducer(state, action)
    lastRenderedState: "hello"
    pending: null
    ▶ __proto__: Object
```

图 4.3

基于 Hook 的数据结构，useEffect/useLayoutEffect 来说，React 主要做的事情分为如下几步：

● render 阶段：函数组件开始渲染的时候，创建出对应的 Hook 链表挂载到 workInProgress 的 memoizedState 上，并创建 effect 链表，也就是挂载到对应的 fiber 节点上，但是基于前一次和本次依赖项的比较结果，创建的 effect 是有差异的。这一点暂且可以理解为：依赖项有变化，effect 可以被处理，否则不会被处理。

● commit 阶段：异步调度 useEffect 或者同步处理 useLayoutEffect 的 effect。等到 commit 阶段完成，更新应用到页面上之后，开始处理 useEffect 产生的 effect，或是直接处理 commit 阶段同步执行阻塞页面时更新的 useLayoutEffect 产生的 effect。

可见，useEffect 和 useLayoutEffect 的执行时机不一样，前者被异步调度，当页面渲染完成后再去执行，不会阻塞页面渲染。后者是在 commit 阶段新的 DOM 准备完成，但还未渲染到屏幕之前，同步执行。

而 useEffect 的工作是使 currentlyRenderingFiber 加载当前的 Hook，具体流程就是判断当前 fiber 是否已经存在 Hook（就是判断 fiber.memoizedState），存在就创建一个 effect Hook 到链表的最后，也就是.next，不存在就创建一个 memoizedState。

再来看一下 Effect 的入口函数：

```
function mountEffect(
  create: () => (() => void) | void,
  deps: Array<mixed> | void | null
): void {
  return mountEffectImpl(
    UpdateEffect | PassiveEffect,
    HookPassive,
    create,
    deps,
  );
};
```

本质上，这一过程涉及调用 mountEffectImpl 函数，其中包含了前面提到的 Effect 类型的相关字段。不过，有一个值得注意的地方是，destroy 函数并没有被传递。事实上，destroy 函数是在上一次 effect 调用中返回的，即在 useEffect 回调中通过 return 返回。然而，由于其在组件的创建阶段是第一次调用，因此当时 destroy 函数为 undefined。

这里看一下创建阶段调用的 mountEffectImpl 函数：

```
function mountEffectImpl(fiberFlags, hookFlags, create, deps): void {
  //创建 Hook 对象
  const hook = mountWorkInProgressHook();
  //获取依赖
  const nextDeps = deps === undefined ? null : deps;
  //为 fiber 挂载上副作用的 effectTag
  currentlyRenderingFiber.flags |= fiberFlags;
  //创建 effect 链表，挂载到 Hook 的 memoizedState 上和 fiber 的 updateQueue 上
  hook.memoizedState = pushEffect(
    HookHasEffect | hookFlags,
    create,
    undefined,
    nextDeps,
  );
};
```

众所周知，useEffect 或 useLayoutEffect 都会根据 deps 的变化重新执行，所以可以猜测，在更新时调用的是 updateEffectImpl 函数。其对比 mountEffectImpl 函数多出来的一部分内容其实就是对比上一次的 effect 的依赖变化，以及执行上一次 effect 中的 destory 部分内容，代码如下：

```
function updateEffectImpl(fiberFlags, hookFlags, create, deps): void {
  const hook = updateWorkInProgressHook();
```

```
  const nextDeps = deps === undefined ? null : deps;
  let destroy = undefined;

  if (currentHook !== null) {
    // 从 currentHook 中获取上一次的 effect
    const prevEffect = currentHook.memoizedState;
    // 获取上一次 effect 的 destory 函数，即 useEffect 回调中返回的函数
    destroy = prevEffect.destroy;
    if (nextDeps !== null) {
      const prevDeps = prevEffect.deps;
      // 比较前后依赖，push 一个不带 HookHasEffect 的 effect
      if (areHookInputsEqual(nextDeps, prevDeps)) {
        pushEffect(hookFlags, create, destroy, nextDeps);
        return;
      }
    }
  }

  currentlyRenderingFiber.flags |= fiberFlags;
  // 如果前后依赖有变，在 effect 的 tag 中加入 HookHasEffect
  // 并将新的 effect 更新到 hook.memoizedState 上
  hook.memoizedState = pushEffect(
    HookHasEffect | hookFlags,
    create,
    destroy,
    nextDeps,
  );
};
```

可以看到在 mountEffectImpl 和 updateEffectImpl 中，最后的结果走向都是 pushEffect 函数，它的工作很纯粹，就是创建出 effect 对象，把对象挂到链表中。

pushEffect 代码如下：

```
function pushEffect(tag, create, destroy, deps) {
  // 创建 effect 对象
  const effect: Effect = {
    tag,
    create,
    destroy,
    deps,
    next: (null: any),
  };
```

```
    // 从 workInProgress 节点上获取到 updateQueue, 为构建链表做准备
    let componentUpdateQueue: null | FunctionComponentUpdateQueue =
(currentlyRenderingFiber.updateQueue: any);
    if (componentUpdateQueue === null) {
    // 如果 updateQueue 为空，把 effect 放到链表中，和它自己形成闭环
      componentUpdateQueue = createFunctionComponentUpdateQueue();
      // 将 updateQueue 赋值给 WIP 节点的 updateQueue，实现 effect 链表的挂载
      currentlyRenderingFiber.updateQueue = (componentUpdateQueue: any);
      componentUpdateQueue.lastEffect = effect.next = effect;
    } else {
      // updateQueue 不为空，将 effect 接到链表的后边
      const lastEffect = componentUpdateQueue.lastEffect;
      if (lastEffect === null) {
        componentUpdateQueue.lastEffect = effect.next = effect;
      } else {
        const firstEffect = lastEffect.next;
        lastEffect.next = effect;
        effect.next = firstEffect;
        componentUpdateQueue.lastEffect = effect;
      }
    }
    return effect;
};
```

这里的主要逻辑其实就是本节开头所说的"区分两种情况"，即链表为空或链表存在的情况。值得一提的是这里的 updateQueue 是一个循环链表。

以上就是 effect 链表的构建过程。可见 effect 对象创建出来最终会以两种形式放到两个地方：

- 单个 effect，放到 hook.memorizedState 上；
- 环形的 effect 链表，放到 fiber 节点的 updateQueue 中。

两者各有用途，前者的 effect 会作为上次更新的 effect，为本次创建 effect 对象提供参照（对比依赖项数组），后者的 effect 链表会作为最终被执行的主体，带到 commit 阶段处理。

在 React 的更新流程中，当完成更新并进入提交阶段时，主要由 commitRoot 函数负责执行。Effect 的创建与其插入到 Hook 链表的过程是在这之前进行的。尽管 useEffect 和 useLayoutEffect 在这方面的处理相同且共享代码，但在提交阶段，它们的执行时机有所不同。这个实现细节可以在 ReactFiberWorkLoop.js 文件中找到，代码如下：

```
    // src/react-reconciler/src/ReactFiberWorkLoop.js
```

```
function commitRoot(root) {
  // 已经完成构建的 fiber，上面会包括 Hook 信息
  const { finishedWork } = root;

  // 如果存在 useEffect 或者 useLayoutEffect
  if ((finishedWork.flags & Passive) !== NoFlags) {
    if (!rootDoesHavePassiveEffect) {
      rootDoesHavePassiveEffect = true;
      // 开启下一个宏任务
      requestIdleCallback(flushPassiveEffect);
    }
  }

  console.log('start commit.');
  // 判断自己身上有没有副作用
  const rootHasEffect = (finishedWork.flags & MutationMask) !== NoFlags;
  // 如果自己有副作用或者子节点有副作用就进行 DOM 操作
  if (rootHasEffect) {
    console.log('DOM 执行完毕');
  commitMutationEffectsOnFiber(finishedWork, root);
  // 执行 layoutEffect
  console.log('开始执行 layoutEffect');
    commitLayoutEffects(finishedWork, root);
    if (rootDoesHavePassiveEffect) {
      rootDoesHavePassiveEffect = false;
      rootWithPendingPassiveEffects = root;
    }
  }
  // 等 DOM 变更后，更改 root 中 current 的指向
  root.current = finishedWork;
};
```

这里的 rootDoesHavePassiveEffect 是核心判断点。还记得 Effect 类型中的 tag 参数吗？就是依靠这个参数来标识区分 useEffect 和 useLayoutEffect 的。

当 rootDoesHavePassiveEffect 为 false 时，副作用函数会被推入宏任务队列。可以理解为 useEffect 的回调被包装在 requestIdleCallback 中异步执行，浏览器将根据当前帧是否有空闲时间来决定副作用函数的具体执行时机。

接着，如果 rootHasEffect 为 true，则说明有副作用需要处理。副作用函数如果是 useEffect，则已经在宏任务队列中排队了；而副作用函数如果是 useLayoutEffect，则会立即执行副作用。这意味着，useEffect 的副作用是异步执行的，不会阻塞页面更新；而 useLayoutEffect 的副作用是同步执行的，会阻塞页面更新，因此不适

合放入复杂逻辑。

4.6 useRef

useRef 是 React 16.8 新增的一个内置 Hook，用于引用一个不需要渲染的值。举个例子：

```
import { useRef } from 'react';
const ref = useRef(0);
ref.current === 0;
```

总结 useRef 的几个特点：

- useRef 返回一个可变的 ref 对象，该对象只有一个 .current 属性，初始值为传入的参数化（initialValue）；
- 返回的 ref 对象在组件整个生命周期内保持不变；
- 当更新 current 值时并不会重新渲染（re-render），这是与 useState 最不同的地方。

通俗一点来说，useRef 就像是可以在 .current 属性中保存一个可变值的盒子。看到 useRef 的 ref 肯定会想到类组件中的 createRef API，它们大部分用法都基本一致，都可以存储变量或者 DOM 节点。

可以通过 ref.current 属性访问这个 ref 的当前变量，这个变量是有意设置为可变的，这意味着可以对它进行读写操作。

例如：

```
const StopWatch = () => {
  const [now, setNow] = useState(Date.now())
  const ref = useRef()

  const handleStart = useCallback(() => {
    ref.current = setInterval(() => {
      setNow(Date.now())
    }, 1000)
  }, [])

  const handleStop = useCallback(() => {
    clearInterval(ref.current)
  }, [])

  return (
```

```
    <>
      <h1>Now Time : {now}</h1>
      <button onClick={handleStart}>Start</button>
      <button onClick={handleStop}>Stop</button>
    </>
  )
};
```

在上面的例子中,我们使用 ref 存储了 setInterval 返回的 ID,需要清除时,开发者只需要清除 ref 即可。如果不用 ref,使用普通变量,则会频繁出现由 state 或 props 导致的组件更新,每一次定时器的 ID 都不一样,并且定时器没有被清除,出现漏洞。

再看一个例子:

```
const ClickWatch = () => {
  let ref = useRef(0);

  const handleClick = useCallback(() => {
    ref.current = ref.current + 1;
    alert('You clicked ' + ref.current + ' times!');
  }, [])

  return (
    <button onClick={handleClick}>
      Click me!
    </button>
  );
};
```

在这个例子中,按钮每次被点击时,ref.current 都会增加,ref 指向了一个数字。和 state 一样,ref 可以指向任何东西,比如字符串、对象、函数。与 state 不同的是,ref 是一个带有当前属性的普通 JavaScript 对象,可以读取和修改。

注意:组件不会在每次 ref 递增时重新渲染,和 state 一样,ref 在每次重新渲染之间会被 React 保留。然而,setState 将会重新渲染组件,但设置 ref 则不会。

useRef 还可以获取 React JSX 中的 DOM 元素,获取后就可以控制 DOM 实体,就像这样:

```
const TextInputWithFocusButton = () => {
  const inputEl = useRef()

  const handleFocus = useCallback(() =>
```

```
    // `current` 指向已挂载到 DOM 上的文本输入元素
    inputEl.current.focus()
  }, [])

  return (
    <div>
      <input ref={inputEl} type="text" />
      <button onClick={handleFocus}>Focus the input</button>
    </div>
  )
};
```

那么什么时候需要使用 useRef 呢？通常，当组件需要跳出 React 与外部 API（原生 API）通信时，就需要用到 ref，即使用不会影响组件外观的浏览器 API，例如下面这些情况：

- 存储 timeout/interval ID；
- 存储和操作 DOM；
- 存储不需要计算 JSX 的其他对象。

另外，如果你的组件需要存储一些值，但是这些值不会影响到渲染的逻辑，就可以选择使用 ref 来存储。那么什么是 ref 的最佳实践呢？通常会有如下几点，遵循下面的原则会使组件和 React 应用更有可预测性。

- 将 ref 视为逃生舱：当使用外部浏览器 API 时，ref 是一个很有用的方式。如果大部分应用程序逻辑和数据流依赖于 ref，可能需要重新考虑一下方法是否正确，因为 ref 的可变性可能会使逻辑和数据流变得不好预测。
- 不在渲染期间读取或写入 ref.current：如果在渲染过程中需要一些数据，请不要使用 ref，而是改用 state。因为 React 不知道 ref.current 是何时发生变化的，在渲染时读取它会使得组件的行为难以预测，并且 React 本身引发视图重渲染也是基于 state 发生变化的。

state 就像渲染后的快照，改变后并不会同步更新。但是 ref 不一样，当改变 ref 时，它会立即改变（同步修改）。ref 本身是一个 JavaScript 对象。在 React 组件中，由于经历了多次无关渲染，ref 会持续保存在组件的内存中，帮助执行业务逻辑。因此，使用 ref 时，不必担心意外变化引发的错误。只要不将这个正在变更的对象用于渲染，React 就不会在意对 ref 的任何操作。

useRef 和 useState 的区别如表 4.1 所示。

下例为一个使用 useState 实现的计数器按钮：

```
const Counter = () => {
  const [count, setCount] = useState(0);
```

```
  const handleClick = () => {
    setCount(count + 1);
  }

  return (
    <button onClick={handleClick}>
      You clicked {count} times
    </button>
  );
};
```

表 4.1　useRef 和 useState 的区别

useRef	useState
useRef(initialValue)返回{ current: initialValue }	useState(initialValue)返回 state 的当前值和一个状态设置函数([value, setState])
改变时不会触发重新渲染	改变时触发重新渲染
可变,在 rendering 过程之外修改和更新 current 的值	不可变,必须使用状态更改函数修改,进队列重新渲染
不应该在 rendering 过程中读或写 current 的值	可以在任何时间读取 state,每次改变都会造成重新渲染

因为显示了 count 的值,所以使用 state 是有意义的。通过 setCount()设置计数器的值时,React 会重新渲染组件并更新屏幕以展示新的 count。

但如果试图用 useRef 来实现类似的计数器按钮,代码如下:

```
const Counter = () => {
  const countRef = useRef(0);
  const handleClick = () => {
    //这里不会重新渲染这个组件!
    countRef.current = countRef.current + 1;
  }

  return (
    <button onClick={handleClick}>
      You clicked {countRef.current} times
    </button>
  );
};
```

尽管该代码可能看起来没什么问题,但是 React 永远都不会重新渲染组件(只会渲染一次),因此看不到计数变化。不管怎么点击按钮,页面始终会显示"You clicked 0 times"。因为页面只执行了一次 mount 阶段的渲染,在那之后,没有触发重新渲染的条件。

在了解了 useRef 之后，你可能会回想起此前曾提及的类组件 createRef。它与 useRef 有何关系？让我们通过一个案例来理解其中区别。

```
const useRefAndCreateRef = () => {
  const [count, setCount] = useState(1)
  const refFromUseRef = useRef()
  const refFromCreateRef = createRef()

  if (!refFromUseRef.current) {
    refFromUseRef.current = count
  }

  if (!refFromCreateRef.current) {
    refFromCreateRef.current = count
  }

  const handleClick = () => {
    setCount((prev) => prev + 1)
  }

  return (
    <div>
      <p>count: {count}</p>
      <p>refFromUseRef{refFromUseRef.current}</p>
      <p>refFromCreateRef{refFromCreateRef.current}</p>
      <button onClick={handleClick}>Cause re-render</button>
    </div>
  )
};
```

点击按钮后组件重新渲染，但是由于 useRef 创建的 ref 一直存在，使用的是同一个引用，无法重新赋值，所以一直都是 1；而 createRef 创建的 ref 每次都会返回一个新的引用，每次更新 count 值都会将 ref 最新值赋值给 createRef。因此得出结论：createRef 每次渲染都会返回一个新的引用，而 useRef 每次都会返回相同的引用。

4.7　useMemo 和 useCallback

useMemo 和 useCallback 这两个 Hook 非常相似，在 React 中的实现原理和用法以及使用场景都是一致的，都是用于主动性能优化的 Hook，它们的主要功能与区别如下：

- useMemo 用于缓存计算结果，只有在依赖项改变时才会重新计算并返回新的值，否则直接返回缓存的值。useMemo 接收两个参数：一个函数和一个依赖数组。
- useCallback 用于返回一个函数，其会在依赖项改变时返回一个新的函数引用。

先看一下 useMemo 的例子：

```
const ExpensiveComponent = ({a, b}) => {
  const result = useMemo(() => {
    console.log("Expensive calculation...");
    return a * b;
  }, [a, b]);

  return <div>Result: {result}</div>;
};
```

我们创建了一个名为 ExpensiveComponent 的组件，它接收两个 props：a 和 b。在组件内部，定义了一个 result 函数，并使用 useMemo 对计算 a 和 b 乘积的过程进行了优化。通过这种方式，只有当 a 或 b 发生变化时，result 函数内部的逻辑才会被重新执行，避免了不必要的计算。这种主动优化的策略特别适合用于具有大量函数状态的组件，能够有效提高性能并减少处理复杂计算时的开销。

对比基于 useCallback 的例子：

```
const ButtonComponent = ({ onClick, children }) => {
  return <button onClick={onClick}>{children}</button>;
};

const ParentComponent = () => {
  const handleClick = useCallback(() => {
    console.log("Button clicked");
  }, []);
  return (
    <div>
      <ButtonComponent onClick={handleClick}>Click me</ButtonComponent>
    </div>
  );
};
```

在这个案例中，我们创建了一个名为 ButtonComponent 的组件，接收来自父组件的 onClick props。子组件 props 不改变，则不会重新渲染。因此当 ParentComponent 重新渲染时，useCallback 会返回上一次创建的内存中的 handleClick 函数实例，避免了不必要的函数创建和组件重渲染。

接下来，我们深入 useMemo、useCallback 的源码进行分析。首先在 React Hooks 体系中，useMemo、useCallback 跟其他 Hook 一样，都分为 mount 和 update 阶段，也就是每个 Hook 都有自己各个阶段的执行逻辑，存放在对应的 Dispatcher 中。以 useMemo 为例：

```
// 挂载时的调度器
const HooksDispatcherOnMount: Dispatcher = {
  // useMemo 挂载时的执行函数
  useMemo: mountMemo,
  // 其他 hook...
};
// 数据更新时的调度器
const HooksDispatcherOnUpdate: Dispatcher = {
  // useMemo 挂载时的执行函数
  useMemo: updateMemo,
  // 其他 Hook...
};
// 其他生命周期调度器...
```

从上面的代码可以看出，useMemo 在挂载阶段执行了 mountMemo，在更新阶段执行了 updateMemo。updateMemo 源码在 packages/react-reconciler/src/ ReactFiberHooks.js 中可以找到。在 updateMemo 的实现中，有一个关键函数 areHookInputsEqual，它用于比较依赖项数组。

areHookInputsEqual 函数接收两个依赖项数组 nextDeps 和 prevDeps。它首先会检查两个数组的长度是否相等，如果不相等，将会在开发环境中发出警告；然后遍历依赖项数组并使用 is 函数（类似于 Object.is）逐个比较元素，如果发现任何不相等的元素，函数将返回 false，否则返回 true。这个函数在 useMemo 的实现中起到了关键作用，因为它决定了是否需要重新计算值。

对于 useCallback 其原理与 useMemo 非常接近，都是通过 areHookInputsEqual 函数进行依赖项对比来决定是返回上一次的值还是重新计算返回新值，区别在于 useMemo 返回新的数据对象，而 useCallback 返回的是回调函数。其源码如下：

```
function updateCallback<T>(callback: T, deps: Array<mixed> | void | null): T {
  const hook = updateWorkInProgressHook();
  const nextDeps = deps === undefined ? null : deps;
  const prevState = hook.memoizedState;
  if (nextDeps !== null) {
    const prevDeps: Array<mixed> | null = prevState[1];
```

```
      if (areHookInputsEqual(nextDeps, prevDeps)) {
        return prevState[0];
      }
    }
    hook.memoizedState = [callback, nextDeps];
    return callback;
};
```

4.8 useContext

2.2.7 节跨组件分层通信中已经提到了 useContext 的应用场景以及源码分析，在本节先回顾一下 useContext 的应用和原理。

使用 useContext 的基本模式如下：

```
const UserNameContext = React.createContext();
function Dev() {
  const userName = React.useContext(UserNameContext);
  return userName.name;
}
function App() {
  return (
    <UserNameContext.Provider value={{name: 'Joe'}}>
      <Dev />
    </ThemeContext>
  );
}
```

通过上述代码可以看到 useContext 的应用场景一共分成三个部分：
- createContext：创建 context 数据源，向下透传；
- Context.Provider：包裹父组件，作为外层组件，传递数据；
- useContext：子孙组件消费 context 的方式。

通过分析代码，可以得到如图 4.4 所示 React Context 的流程图。其中，包含了四个主要角色：
- fiber：React 内部用于描述组件状态和行为的数据结构；
- provider：组件，通过<Context.Provider>将值放入特定的上下文中；
- consumer：组件，通过<Context.Consumer>从上下文中获取值；
- Dev：一个函数式组件，通过 useContext 来订阅上下文的变化并获取值。

这张流程图展示了 React 中的上下文机制是如何通过 Provider、Consumer 和 useContext 来实现数据传递和更新的，具体步骤如下：

图 4.4

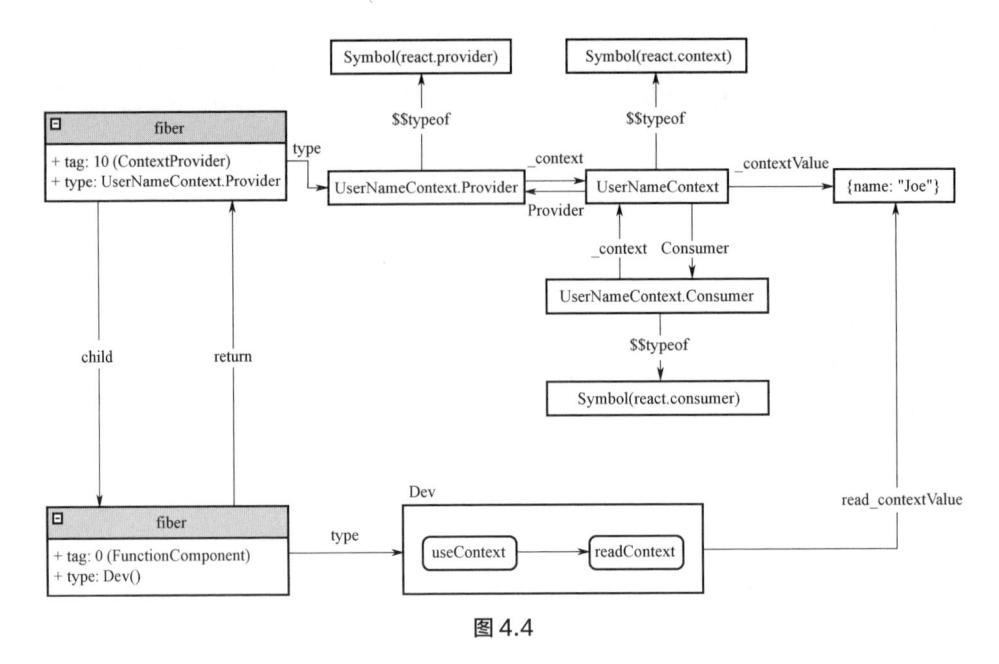

- Context 的创建，UserNameContext 是通过 React.createContext 创建的，生成了一个拥有特定属性的对象，如$$typeof（标记这个对象是一个 Context），以及 Provider 和 Consumer 对象。
- Provider 的使用，UserNameContext.Provider 组件接收一个名为 name 的属性，这个属性值会作为 Context 初始值保存在 UserNameContext.Provider 中。
- Consumer 的使用，当 Consumer 组件需要访问上下文时，它会以 UserNameContext.Consumer 对象中查找_contextValue 字段，读取当前的上下文。
- 函数组件的上下文使用，在函数组件 Dev 中，使用 useContext 钩子来订阅 UserNameContext，useContext 会读取上下的值，并在值发生变化时触发组件重新渲染。

使用 useContext 时返回的_currentValue 提供给使用的组件，通过源码分析发现，每次读取 context 都会向 lastContextDependency 推送数据。每次 Context.Provider 的 beginWork 都会向 valueStack 的栈中推送数据，在 Context.Provider 的 CompleteUnitOfWork 的时候，会从 valueStack 的栈里抛出数据，需要注意的是：

- 最新的 context 值在这个数据结构中没有存储；
- 最新的 context 值之前存储在 valueCursor.current 中；
- 除此之外的值才会存储到 valueStack 中。

举例来讲，当第一次使用 Context.Provider 时，valueStack 如图 4.5 所示。
有三层嵌套的 Context.Provider 时，valueStack 如图 4.6 所示。

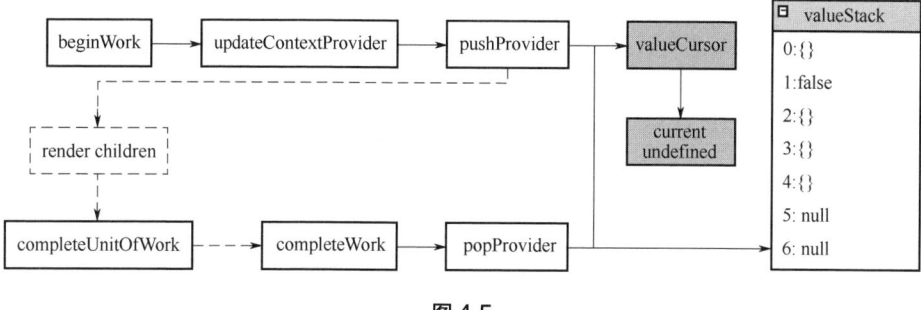

图 4.5

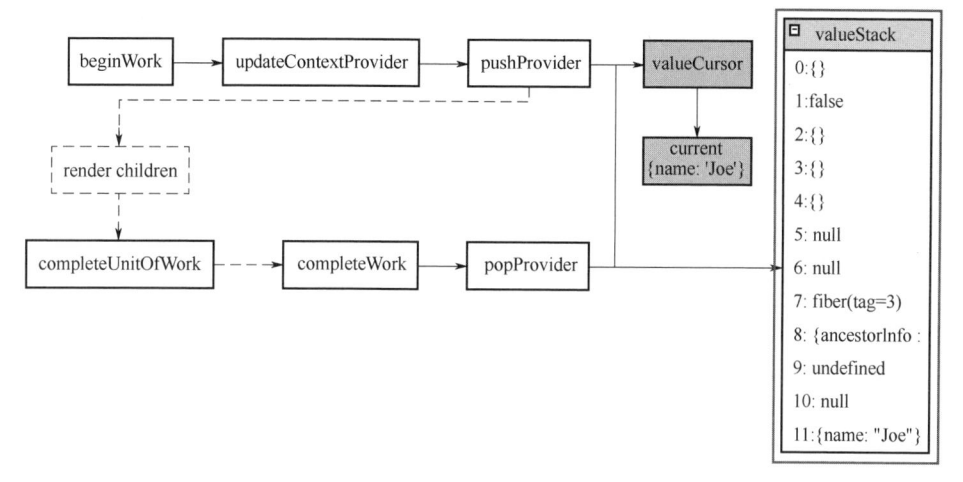

图 4.6

对应的参考代码如下，一共三层嵌套关系：

```
const UserNameContext = React.createContext();
const Provider = UserNameContext.Provider;

function Deep2() {
const userName = React.useContext(UserNameContext);
return userName.name;
}


function Deep() {
const userName = React.useContext(UserNameContext);
return (<Provider value={userName}>
<Deep2 />
</Provider>);
}
```

```
function Dev() {
  const userName = React.useContext(UserNameContext);
  return (<div id="div">
  <div>{userName.name}</div>
  <Provider value={userName}>
  <Deep />
  </Provider>
  </div>);
}

function App() {
  const [name, setName] = React.useState('Joe');
  return (
  <Provider value={{ name: name }}>
  <Dev />
  </Provider>
  );
}
```

更加具体的 useContext 实现原理可回顾 2.2.7 节跨组件分层通信。

4.9 useReducer

众所周知，useState 经常用于单个组件的状态管理，但是遇到复杂状态时，useState 可能无法满足需求。而 useReducer 就是为了解决这类复杂需求而生的。从使用层面来说，useReducer 是可以代替 useState 的。这里有个定律：组件内部状态采用 useState 管理；项目工程的状态采用 useReducer 管理。

useReducer 和 useState 最根本的区别在于，前者通过 action 来更新状态，后者直接通过改值的形式来更新状态。显然，useReducer 的更新方式会使更新逻辑更有可读性和可维护性。useReducer 的用法举例如下：

```
const [state, dispatch] = useReducer(reducer, initialArg, init?)
```

通常情况下，只会用到 useReducer 的前两个参数，如下面这个计数器组件：

```
const initialState = { count: 0 };

function reducer(state, action) {
  switch (action.type) {
    case 'increment':
      return { count: state.count + 1 };
```

```
    case 'decrement':
      return { count: state.count - 1 };
    default:
      throw new Error();
  }
}

function Counter() {
  const [state, dispatch] = useReducer(reducer, initialState);

  return (
    <>
      Count: {state.count}
      <button onClick={() => dispatch({ type: 'decrement' })}>-</button>
      <button onClick={() => dispatch({ type: 'increment' })}>+</button>
    </>
  );
};
```

这段代码中声明了四个变量：
- state：提供给组件调用的状态，与 useState 返回数组第一个值作用一样；
- dispatch：用于调用 reducer 中操作的方法；
- reducer：声明当前状态可操作的所有动作（action）；
- initialState：声明初始状态。

使用 dispatch 的注意事项：当 dispatch 调用后，状态更新是异步的，并不会立即更新状态，这一点和 useState 的更新机制是一样的。并且 React 对 dispatch 有一个优化机制，当 dispatch 触发前后的值相同（源码中通过 Object.is()判断）时 React 就不会进行重新渲染，这是一个小的性能优化点。

```
function addCount() {
  dispatch({ type: 'increment' })
  console.log(state.count)  // 打印出来的不是新值
}

<button onClick={addCount}>+</button>
```

使用 reducer 的注意事项：首先 reducer 是更新动作，动作中都是更新对象和数组，对于 Object 类型，需要创建一个新的对象或数组，而不是在原数组上进行修改，否则将不会渲染，这点也是和 useState 一样的。

接下来是 useReducer 的初始化，即 useReducer 的第三个参数 init。那么它的作用是什么呢？其实也是为了性能优化。假设一个场景，如果计数器组件缓存在 localStorage 里，当进入页面的时候，我们希望从 localStorage 中读取值来作为 useReducer 的初值，如果没有 init 的话，可能会这样写：

```js
function getInitialCount() {
  const savedCount = localStorage.getItem("count");
  return savedCount ? Number(savedCount) : 0;
}

function counterReducer(state, action) {
  switch (action.type) {
    case "INCREMENT":
      return { count: state.count + 1 };
    case "DECREMENT":
      return { count: state.count - 1 };
    default:
      return state;
  }
}

function Counter() {
  const [state, dispatch] = useReducer(counterReducer, getInitialCount());
  // 使用 useEffect 来监听状态的变化，并将其保存到 localStorage
  useEffect(() => {
    localStorage.setItem("count", state.count);
  }, [state.count]);
  return (
    <>
      Count: {state.count}
      <button onClick={() => dispatch({ type: "INCREMENT" })}>+1</button>
      <button onClick={() => dispatch({ type: "DECREMENT" })}>-1</button>
    </>
  );
}
```

在这个例子中，我们直接调用了 getInitialCount 函数并将返回值作为 useReducer 的第二个参数，从而初始化状态。当 React 初始化这个组件时，它就会执行这个函数并使用其返回值作为初始状态。

若使用第三个参数 init 进行初始化，代码可以改成这样：

```js
function init(initialValue) {
  // 尝试从 localStorage 中读取值
```

```jsx
  const savedCount = localStorage.getItem("count");
  // 如果有值并且可以被解析为数字, 则返回它, 否则返回 initialValue
  return { count: savedCount ? Number(savedCount) : initialValue };
}

function counterReducer(state, action) {
  switch (action.type) {
    case "INCREMENT":
      return { count: state.count + 1 };
    case "DECREMENT":
      return { count: state.count - 1 };
    default:
      return state;
  }
}

function Counter() {
  const [state, dispatch] = useReducer(counterReducer, 0, init);
  // 使用 useEffect 来监听状态的变化, 并将其保存到 localStorage
  useEffect(() => {
    localStorage.setItem("count", state.count);
  }, [state.count]);
  return (
    <>
      Count: {state.count}
      <button onClick={() => dispatch({ type: "INCREMENT" })}>
        +1
      </button>
      <button onClick={() => dispatch({ type: "DECREMENT" })}>
        -1
      </button>
    </>
  );
}
```

这两种方式看起来差不多, 但它们的区别其实很大, 可以分为四点:

① 执行时机:
- 直接调用函数作为第二个参数, 这个函数在组件每一次渲染时都会执行。
- 使用 init 函数, 只会在组件初始化时渲染一次。

② 访问到的数据:
- 直接调用函数作为第二个参数, 这个函数只能访问到定义它时作用域内的

数据。
- 使用 init 函数，由于 init 函数接收 initialArg 作为参数，因此 init 函数具有更大的灵活性，能够基于入参去计算。

③ 代码：
- 直接调用函数作为第二个参数，代码更简洁，适用于简单初始化逻辑。
- 使用 init 函数，提供了更清晰的代码组织结构，特别是初始化逻辑很复杂的时候，单独提炼出一个函数会很有必要。

④ 性能：
- 直接调用函数作为第二个参数，如果函数执行了一些计算密集的操作，那么状态每次更新都会导致组件重渲染，从而重新初始化一遍，很浪费性能。
- 使用 init 函数，只在初始化阶段执行一次，这样会更高效。

总结一下，两者都可以用于初始化 reducer 的初始状态，如果初始化逻辑很简单，那其实两者没什么差别，直接用函数作为第二个参数也无所谓；如果逻辑较为复杂，那么应该使用 init 函数，以保障性能。

useContext 的最核心技术就是透传组件状态，但其对于透传到子组件的状态，只能进行展示，并不能向根层更改，因此可以通过 useContext+useReducer 来实现一个迷你的状态管理机，根组件把状态和 reducer 一起透传出去，从而实现子孙组件 action 改变父组件的 context。

我们来实现一个主题切换系统，首先我们定义 state、context 和 reducer：

```
const ThemeContext = React.createContext();
const initialState = { theme: 'light' };
function themeReducer(state, action) {
  switch (action.type) {
    case 'TOGGLE_THEME':
      return { theme: state.theme === 'light' ? 'dark' : 'light' };
    default:
      return state;
  }
}
```

然后创建 Provider 组件：

```
function ThemeProvider({ children }) {
const [state, dispatch] = useReducer(themeReducer, initialState);
return (
    <ThemeContext.Provider value={{ theme: state.theme, toggleTheme: () => dispatch({ type: 'TOGGLE_THEME' }) }}>
      {children}
```

```
      </ThemeContext.Provider>
  );
}
```

在透传的子组件中,可以随意用 dispatch 处理来自父组件的 context,因为父组件把对应的 dispatch 也作为 context 传递下去了,代码如下:

```
function ThemedButton() {
  const { theme, toggleTheme } = useContext(ThemeContext);
  return (
    <button style={{ backgroundColor: theme === 'light' ? '#fff' : '#000' }}
      onClick={toggleTheme}>
      Toggle Theme
    </button>
  );
}
```

这样,一个迷你 Redux 状态管理闭环就实现了。虽然 useReducer 和 Redux 都采用了 action 和 reducer 的模式来处理状态,但是它们的实现和使用上还是会有区别的。

● 范围:useReducer 通常在组件或小型应用中使用,而 Redux 被设计为大型应用的全局状态管理工具。

● 中间件和扩展:Redux 支持中间件,这允许开发者插入自定义逻辑,例如日志、异步操作等;而 useReducer 本身不直接支持中间件,但我们可以模拟中间件的效果。

● 复杂性:对于简单的状态管理,useReducer 通常更简单和直接。但当涉及复杂的状态逻辑和中间件时,Redux 可能更具优势。

4.10 自定义 Hook

在 React 项目中,开发者经常会使用到 React 自带的几个内置 Hook,但在有些场景下,可能还需要一些自定义的 Hook 以满足特定需求,例如获取数据、获取链接、发出网络请求等。虽然 React 没有提供偏业务性的 Hook,但是基于内置 Hook 以及框架特性,我们可以根据需求自定义 Hook。

关于 Hook 的命名规范,Hook 名必须以 use 开头,使用驼峰命名法清晰描述功能,如 useDate、useConnect、useRequest,此外还应避免与内置 Hook 重名。

自定义 Hook 的核心是共享组件之间的逻辑。自定义 Hook 不仅可以减少重复代码,更重要的是,它通过封装具体的实现细节,让使用者更关注功能本身,而不是实现方式。当开发者将逻辑提取到自定义 Hook 中时，可以隐藏处理某些"外部系统"或浏览器 API 调用的细节。下面是一个简单的例子：

```
import { useState } from 'react';

const useCounter = (initialValue) => {
  const [count, setCount] = useState(initialValue);
  function increment() {
    setCount(count + 1);
  }
  return [count, increment];
}
```

这是一个用于计数的 Hook——useCounter，其入参为一个初始 count 值，返回更新后的 count 和增加 count 的方法。如果开发者有点击按钮后记录点击次数的需求，那么可以像这样去消费：

```
import React from 'react';
import useCounter from './useCounter';

const Counter = () => {
  const [count, increment] = useCounter(0);
  return (
    <div>
      <p>Count: {count}</p>
      <button onClick={increment}>Increment</button>
    </div>
  );
};
```

在这个例子中，我们导入了 useCounter，并在组件顶层作用域调用，同时将返回的数组解构为 count 和 increment，然后在组件中调用。

同样，自定义 Hook 也支持状态独立，不会由于调用多个相同的 Hook 引发状态共享的问题，即每个 Hook 的调用都完全独立于对同一个 Hook 的其他调用。以 useCounter 为例：

```
import useCounter from './useCounter';

const Counter = () => {
  const [count1, increment1] = useCounter(0);
  const [count2, increment2] = useCounter(100);
  return (
    <div>
```

```
    <p>Count1: {count1}</p>
    <button onClick={increment1}>Increment1</button>
    <p>Count2: {count2}</p>
    <button onClick={increment2}>Increment2</button>
  </div>
 );
};
```

在下面的内容中,我们将列举一些功能型 Hook 和业务型 Hook 的实现及用法。

4.10.1 功能型 Hook

功能型 Hook 是以实现特定功能或目的而生的 Hook,与具体业务无直接关系。以下列举一些比较常见的功能型 Hook。

(1) useWindowWidth

这是一个 React Hook,名为 useWindowWidth,它可以实时返回当前浏览器窗口的宽度,基于 useState 和 useEffect 实现。

```
import { useState, useEffect } from 'react';

const useWindowWidth = () => {
  const [windowWidth, setWindowWidth] = useState(window.innerWidth);

  useEffect(() => {
    const handleResize = () => setWindowWidth(window.innerWidth);
    window.addEventListener('resize', handleResize);
    return () => window.removeEventListener('resize', handleResize);
  }, []);
  return windowWidth;
};
```

上述代码中首先声明了窗口宽度的状态,初始值即页面渲染时的 window.innerWidth,同时开启 useEffect 副作用,监听 window 中的 resize 事件,当事件触发后,异步更新 windowWidth 的状态,执行 setWindowWidth,最后返回该状态即可。useWindowWidth 的使用与常规 Hook 一样,如果需要获取实时的窗口宽度值的话,直接在组件中使用即可。其他的窗口信息都是同理来实现的。

(2) useLocalStorage

这是一个 React Hook,名为 useLocalStorage,它实现了从浏览器的本地存储中获取数据、修改数据,并保持 React 视图同步,基于 useState 实现。

```
import { useState } from 'react';
```

```
const useLocalStorage = (key, initialValue) => {
  const [storedValue, setStoredValue] = useState(() => {
    try {
      const item = window.localStorage.getItem(key);
      return item ? JSON.parse(item) : initialValue;
    } catch (error) {
      console.log(error);
      return initialValue;
    }
  });

  const setValue = (value) => {
    try {
      setStoredValue(value);
      window.localStorage.setItem(key, JSON.stringify(value));
    } catch (error) {
      console.log(error);
    }
  };

  return [storedValue, setValue];
};
```

useLocalStorage 的入参分别是本地存储的 key 值和 value 值，返回结果是 value 值和再次更新 key 的方法。storedValue 的初始值首先会在 localStorage 中查找入参的 key，如果有，则直接返回该 key 的 value；如果没有，则使用入参的第二个参数 value，创建出一个新的本地存储对象。最后将修改该本地存储的方法和最新的值返回到组件中。

使用 Hook 的特点是其状态隔离，这意味着可以在一个组件中分别处理不同的状态。这使得它非常适合处理需要在同一组件中管理多个独立状态的情况。例如，开发者可以使用不同的 Hook 来处理不同的数据请求或用户交互，而不会混淆它们的状态。这种特性使 Hook 在处理复杂的应用程序逻辑时非常有用。

```
import { useEffect } from 'react';

const Component = () => {
  const [name, setName] = useLocalStorage('name', 'mark');
  const [age, setAge] = useLocalStorage('age', 18);

  useEffect(() => {
```

```
    setName('tom');
    setAge(20);
  }, []);

  return (
    <>
      name:{name}
      age:{age}
    </>
  );
};
```

(3) useInterval

useInterval 实现了一个非常有用的改进版定时器函数,它支持立即执行一次函数然后按照设定的时间间隔定期执行,代码如下:

```
import { useRef, useEffect } from 'react';

const useInterval = (fn, delay = 1000, immediate = false) => {
  const fnRef = useRef();
  fnRef.current = fn;

  useEffect(() => {
    if (delay === undefined || delay === null) return;
    if (immediate) {
      fnRef.current?.();
    }
    const timer = setInterval(() => {
      fnRef.current?.();
    }, delay);
    return () => {
      clearInterval(timer);
    };
  }, [delay]);
};
```

useInterval 一共有三个参数:

- fn: 自定义执行函数;
- delay: 自定义时间,默认为 1s;
- immediate: 是否立即执行一次,默认为 false。

该 Hook 首先声明了 fnRef 用于绑定执行事件,在 useEffect 中开启了一个定时器,定时执行函数,最后在 useEffect 的返回值中注册、销毁、清空定时器。使用方式也很简单，就像这样：

```
const Component = () => {
  const log = () => {
    console.log('定时器执行');
  };
  useInterval(log, 1500, true);

  return <></>;
};
```

4.10.2 业务型 Hook

本小节将从实际业务角度出发，介绍几种比较常见的业务型 Hook。

（1）useFetch

实际业务避免不了大量的网络请求，因此封装一个通用的请求 Hook，用以获取请求的数据、加载情况、错误信息等是个不错的想法。我们试着封装一个 useFetch，该 Hook 用于从 API 中获取请求数据，代码如下：

```
import { useState, useEffect } from 'react';

const useFetch = (url) => {
  const [data, setData] = useState(null);
  const [error, setError] = useState(null);
  const [isLoading, setIsLoading] = useState(true);

  useEffect(() => {
    const fetchData = async () => {
      try {
        const response = await fetch(url);
        const json = await response.json();
        setData(json);
      } catch (error) {
        setError(error);
      } finally {
        setIsLoading(false);
      }
    };
    fetchData();
  }, [url]);

  return { data, error, isLoading };
```

```
};
```

useFetch 首先会接收一个服务器接口地址，在 useEffect 中通过原生 fetch 的形式发出网络请求，通过 try catch 的形式监听请求的结果。如果有错误则将 error 信息保存到状态中；如果请求正常，则保存 data 到状态中，并且在整个过程中记录请求的加载状态。最后将结果、错误信息、加载状态返回。

当然，实际的请求会更复杂，还需要支持请求头的配置、请求事件的变更，完整版的请求 Hook 会在 4.10.3 节介绍。

（2）useModal

实际业务中通常会用到很多弹窗，如果是同类型弹窗的话，考虑后期维护，可以基于 Hook 去单独做一个 useModal，代码如下：

```
const useModal = ({ initTitle, initContent }) => {
  const [visible, setVisible] = useState(false);
  const content = initContent;
  const [title, setTitle] = useState(initTitle);

  const CustomModal = (props) => {
    const onOk = () => {
      const { onOk } = props;
      const form = content.props.form;
      onOk(form);
    }
    const onCancel = () => {
      const { onCancel } = props;
      onCancel()
    }

    return (
      <Modal
        visible={visible}
        title={title}
        okText='确定'
        onOk={onOk}
        cancelText='取消'
          onCancel={onCancel}
          maskClosable={false}
      >
        {content}
      </Modal>
    )
```

```
  }

  const show = () => {
    setVisible(true)
  }
  const hide = () => {
    setVisible(false)
  }

  return { show, hide, CustomModal, setTitle }
};
```

useModal 接收两个自定义参数，分别是 initTitle 和 initContent，对应了弹窗的标题和主题内容，同时在 useModal 中声明了 Modal 组件实体，最后将组件实体和 Modal 的状态（包括显示隐藏）都向外暴露出去。同样，由于 content 是自定义的，开发者也可以在 Hook 使用层自定义一些 content 实体的绑定事件，比如插入一个表单进去，这样就做到了低耦合、复用性强的特征了。

useModal 可以像这样去使用：

```
const Component = () => {
  const formContent = () => (
    <Form
      name="basic"
      labelCol={{ span: 8 }}
      wrapperCol={{ span: 16 }}
      initialValues={{ remember: true }}
      onFinish={onFinish}
      onFinishFailed={onFinishFailed}
      autoComplete="off"
    >
      <Form.Item
        label="Username"
        name="username"
        rules={[{required: true, message: 'Please input your username!' }]}
      >
        <Input />
      </Form.Item>

      <Form.Item
        label="Password"
        name="password"
        rules={[{required: true, message: 'Please input your password!' }]}
```

```
      >
        <Input.Password />
      </Form.Item>
      <Form.Item
        name="remember"
        valuePropName="checked"
        wrapperCol={{ offset: 8, span: 16 }}
      >
        <Checkbox>Remember me</Checkbox>
      </Form.Item>
      <Form.Item wrapperCol={{ offset: 8, span: 16 }}>
        <Button type="primary" htmlType="submit">
          Submit
        </Button>
      </Form.Item>
    </Form>
  );

  const { show, hide, Modal, setTitle } = useModal('初始标题', formContent);

  return (
    <>
      <button onClick={show}>点击打开弹窗</button>
    </>
  );
};
```

在上述代码中，弹窗层的所有状态管理都在 useModal 中完成，调用组件时只需要注意渲染逻辑即可，而传入的表单还是由调用端去进行数据录入和交互，这样就把 Hook 聚合在内部，实现复用的效果了。

4.10.3 实现一个完整版 useRequest

在业务逐渐复杂化，一个页面渲染时需要请求很多个接口的时候，会发现使用 async+await 在复杂场景下会很臃肿，出现大量高耦合的代码，使项目难以维护。因此有必要通过一个自定义 Hook（useRequest）实现以下功能：

- 每一个请求单独管理；
- 手动触发请求；
- 轮询和手动取消；
- 依赖请求（串行请求）；

- 防抖和节流；
- 缓存和依赖更新。

本节将通过 React+TypeScript 按照以上顺序一步步完善 useRequest。

① 雏形版本 useRequest。对标 4.10.2 节中的 useFetch，具备响应数据、加载状态、错误信息的捕捉功能，代码如下：

```typescript
interface UseRequestOptionsProps {
  /** 请求参数*/
  initialData?: object;
  /** 请求成功回调*/
  onSuccess?: (res: any) => void;
}

const useRequest = (
  requestFn: (
    initialData?: object | string | [],
  ) => Promise<SetStateAction<any>>,
  options: UseRequestOptionsProps,
) => {
  const [data, setData] = useState<SetStateAction<any>>(null);
  const [loading, setLoading] = useState<boolean>(false);
  const [error, setError] = useState<string | null>(null);
  const { initialData, onSuccess } = options;
  useEffect(() => {
    setLoading(true);
    setError(null);
    setData(null);
    request();
  }, [requestFn]);

  // useRequest 业务逻辑
  const request = async () => {
    try {
      const res = await requestFn(initialData);
      setData(res);
      // 请求成功响应回调
      onSuccess && onSuccess(res);
    } catch (err) {
      err && setError(JSON.stringify(err));
    } finally {
      setLoading(false);
```

```
    }
  };

  return { data, loading, error };
};
```

其在 React 组件中的使用方法如下:

```
const { data, loading, error } = useRequest(
  queryCompensatoryOrderSituation,
  {
    initialData: {compensatoryId,},
    onSuccess: (res) => {
      console.log('success request!', res);
    },
  },
);
```

useRequest 对于请求函数的写法并无过多要求，只要是一个异步 function 并且会返回一个 promise 对象，即可作为第一个参数传入 useRequest 中；而第二个参数则是一系列的可选配置项，当前的雏形版本暂时只支持 onSuccess。

② 在雏形版本基础上增加手动触发的能力，改造后的代码如下:

```
interface UseRequestOptionsProps {
  /** 手动开启*/
  manual?: boolean;
  /** 请求参数*/
  initialData?: object;
  /** 请求成功回调*/
  onSuccess?: (res: any) => void;
}

const useRequest = (
  requestFn: (
    initialData?: object | string | [],
  ) => Promise<SetStateAction<any>>,
  options: UseRequestOptionsProps,
) => {
  const [data, setData] = useState<SetStateAction<any>>(null);
  const [loading, setLoading] = useState<boolean>(false);
  const [error, setError] = useState<string | null>(null);
  const { manual, initialData, onSuccess } = options;
  useEffect(() => {
```

```
    setLoading(true);
    setError(null);
    setData(null);
    !manual && request();
  }, [manual]);

  // useRequest 业务逻辑
  const request = async () => {
    try {
      const res = await requestFn(initialData);
      setData(res);
      // 请求成功响应回调
      onSuccess && onSuccess(res);
    } catch (err) {
      err && setError(JSON.stringify(err));
    } finally {
      setLoading(false);
    }
  };

  return { data, loading, error, request };
};
```

其在雏形版本的基础上，首先增加了 manual 的配置项，代表是否手动开启，默认为 false。在 useEffect 中，如果未配置手动执行，则会在渲染时执行一次 request，发出请求；如果配置了手动执行，则不会执行 request，并且在最后，useRequest 将 request 方法暴露出去，以达到手动执行的需求。

其在组件中的使用方式如下：

```
const { data, loading, error, request } = useRequest(
  queryCompensatoryOrderSituation,
  {
    manual: true,
    initialData: { compensatoryId, },
    onSuccess: (res) => { console.log('success request!', res);},
  },
);

request();
```

总结一下，手动执行的逻辑主要是根据 manual 参数去除 useRequest mount 阶段的渲染请求部分，把执行请求的能力暴露出去，在页面中去手动调用 request()

来触发。

③ 实现轮询和手动取消的能力。在业务场景中，我们经常会遇到轮询的需求，如订单详情页需要一直监听订单状态，当状态改变时更新页面。当然，也可以通过 WebSocket 的方案实现，但是轮询的需求场景还是很多的，并且在轮询到状态变化后，还需要取消轮询，因此基于该需求，我们再来增强一下已有的 useRequest。改造后代码如下：

```
interface UseRequestOptionsProps {
  /** 手动开启*/
  manual?: boolean;
  /** 请求参数*/
  initialData?: object;
  /** 轮询*/
  pollingInterval?: number | null;
  /** 请求成功回调*/
  onSuccess?: (res: any) => void;
}

const useRequest = (
  requestFn: (
    initialData?: object | string | [],
  ) => Promise<SetStateAction<any>>,
  options: UseRequestOptionsProps,
) => {
  const [data, setData] = useState<SetStateAction<any>>(null);
  const [loading, setLoading] = useState<boolean>(false);
  const [error, setError] = useState<string | null>(null);
  const status = useRef<boolean>(false);
  const pollingIntervalTimer = useRef<NodeJS.Timer | null>(null);
  const { manual, initialData, pollingInterval, onSuccess } = options;
  useEffect(() => {
    setLoading(true);
    setError(null);
    setData(null);
    !manual && request();
  }, [manual]);

  // useRequest 业务逻辑
  const request = async () => {
    try {
      !status.current && (status.current = true);
```

```
        if (pollingInterval && status.current) {
          pollingIntervalTimer.current = setTimeout(() => {
            status.current && request();
          }, pollingInterval);
        }
        const res = await requestFn(initialData);
        setData(res);
        // 请求成功响应回调
        onSuccess && onSuccess(res);
      } catch (err) {
        err && setError(JSON.stringify(err));
      } finally {
        setLoading(false);
      }
    };

    return { data, loading, error, request, cancel };
  };

  // 取消
  const cancel = () => {
    if (pollingIntervalTimer.current) {
      clearTimeout(pollingIntervalTimer.current);
      pollingIntervalTimer.current = null;
      status.current && (status.current = false);
    }
  };
```

在轮询的支持实现设计中，首先入参的配置项增加了 pollingInterval，用来表示是否开启轮询模式并且指定轮询间隔时间。当开启轮询模式后，在执行请求的时候就直接通过 setTimeout 的形式去执行了，主要用到了 timer、setTimeout 的递归思路，同时给出一个 status 状态值判断当前请求是否在轮询中：当调用端执行 cancel() 时，status 为 false；当轮询开始后，status 为 true。

而 cancel() 的核心逻辑是取消了 timer 的递归请求逻辑，其实现并不复杂。然后再看一下轮询的业务场景和 manual:true 的配合很多。实现轮询的代码如下：

```
const { data, loading, error, request, cancel } = useRequest(
  queryCompensatoryOrderSituation,
  {
    manual: true,
    initialData: { compensatoryId, },
    pollingInterval: 1000,
```

```
    onSuccess: (res) => {
      console.log('success request!', res);
    },
  },
);
request();
// ...
// 轮询到理想数据后
cancel();
```

④ 实现串行请求，也可以理解为依赖条件式请求。简单来说就是 A 请求依赖于 B 请求，A 请求在 B 请求执行完后才会去发出请求，如图 4.7 所示。

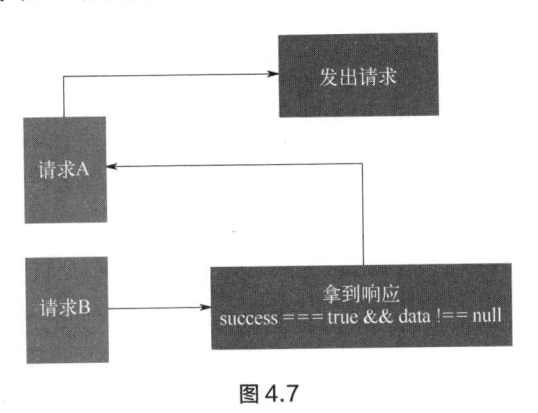

图 4.7

当页面加载后，请求 B 先发出，如果响应结果成功并且有数据，则发出请求 A，否则不发出。那么基于这样的业务场景应该如何改造 useRequest 呢？代码如下：

```
interface UseRequestOptionsProps {
  /** 手动开启*/
  manual?: boolean;
  /** 请求参数*/
  initialData?: object;
  /** 轮询*/
  pollingInterval?: number | null;
  /** 准备，用于依赖请求*/
  ready?: boolean;
  /** 请求成功回调*/
  onSuccess?: (res: any) => void;
}

const useRequest = (
  requestFn: (
```

```
    initialData?: object | string | [],
  ) => Promise<SetStateAction<any>>,
    options: UseRequestOptionsProps,
) => {
  const [data, setData] = useState<SetStateAction<any>>(null);
  const [loading, setLoading] = useState<boolean>(false);
  const [error, setError] = useState<string | null>(null);
  const status = useRef<boolean>(false);
  const pollingIntervalTimer = useRef<NodeJS.Timer | null>(null);
  const {
    manual,
    initialData,
    pollingInterval,
    ready = true,
    onSuccess,
  } = options;
  useEffect(() => {
    setLoading(true);
    setError(null);
    setData(null);
    !manual && ready && request();
  }, [manual, ready]);

  // useRequest 业务逻辑
  const request = async () => {
    try {
      !status.current && (status.current = true);
      if (pollingInterval && status.current) {
        pollingIntervalTimer.current = setTimeout(() => {
          status.current && request();
        }, pollingInterval);
      }
      const res = await requestFn(initialData);
      setData(res);
      // 请求成功响应回调
      onSuccess && onSuccess(res);
    } catch (err) {
      err && setError(JSON.stringify(err));
    } finally {
      setLoading(false);
    }
  };
```

```
  return { data, loading, error, request, cancel };
};

// 取消
const cancel = () => {
  if (pollingIntervalTimer.current) {
    clearTimeout(pollingIntervalTimer.current);
    pollingIntervalTimer.current = null;
    status.current && (status.current = false);
  }
};
```

依赖请求的思路就是在 Hook 中加入一个 ready 字段，也是在基于手动执行 manual 一层的限制后并行再加入一层，来判断是否满足 Hook 发出请求的条件，是否在 Hook 加载时进行默认请求。而当 option 中的 ready 更新（为 true）时，Hook 自动更新从而发出请求。

这一功能常用于页面中请求 B 执行完以后再执行请求 A，A 请求的 ready 字段依赖于 B 请求的 data 或 loading 时，使用的话就可以像这样：

```
const [mountLoading, setMountLoading] = useState<boolean>(false);
useEffect(() => {
  setMountLoading(true);
}, [2000])

const { data, loading, error, request, cancel } = useRequest(
  queryCompensatoryOrderSituation,
  {
    initialData: { compensatoryId,},
    pollingInterval: 1000,
    ready: mountLoading,
    onSuccess: (res) => {
      console.log('success request!', res);
    },
  },
);
```

页面中有一个 mountLoading 状态，页面渲染 2s 后更改状态，useRequest 请求也发送出去了。如果多个请求都采用 useRequest 的形式，由于 Hook 状态相互隔离，不必担心状态污染的情况，只需要采用不同命名的形式即可，就像这样：

```
const { aRequestData,
```

第 4 章 React Hooks 深入浅出　151

```
  aLoading,
  aError,
  aRequest,
  acCancel
} = useRequest(
// ...
);
```

⑤ 实现防抖与节流。由于发出多次请求会很消耗网络资源并影响前端页面性能，因此对于防抖与节流的支持也很重要。其实现思路比较简单，依赖于 lodash 库，包装一下 request 函数的请求实体内容即可，代码如下：

```
interface UseRequestOptionsProps {
  /** 手动开启 */
  manual?: boolean;
  /** 请求参数 */
  initialData?: object;
  /** 轮询 */
  pollingInterval?: number | null;
  /** 准备，用于依赖请求 */
  ready?: boolean;
  /** 防抖 */
  debounceInterval?: number;
  /** 节流 */
  throttleInterval?: number;
  /** 请求成功回调 */
  onSuccess?: (res: any) => void;
}

const useRequest = (
  requestFn: (
    initialData?: object | string | [],
  ) => Promise<SetStateAction<any>>,
  options: UseRequestOptionsProps,
) => {
  const [data, setData] = useState<SetStateAction<any>>(null);
  const [loading, setLoading] = useState<boolean>(false);
  const [error, setError] = useState<string | null>(null);
  const status = useRef<boolean>(false);
  const pollingIntervalTimer = useRef<NodeJS.Timer | null>(null);

  const {
    manual,
```

```js
    initialData,
    pollingInterval,
    ready = true,
    debounceInterval,
    throttleInterval,
    onSuccess,
  } = options;

  useEffect(() => {
    setLoading(true);
    setError(null);
    setData(null);
    !manual && ready && request();
  }, [manual, ready]);

  // 请求
  const request = () => {
    if (debounceInterval) {
      lodash.debounce(requestDoing, debounceInterval)();
    } else if (throttleInterval) {
      lodash.throttle(requestDoing, throttleInterval)();
    } else {
      requestDoing();
    }
  };

  // useRequest 业务逻辑
  const requestDoing = async () => {
    try {
      !status.current && (status.current = true);
      if (pollingInterval && status.current) {
        pollingIntervalTimer.current = setTimeout(() => {
          status.current && request();
        }, pollingInterval);
      }
      const res = await requestFn(initialData);
      setData(res);
      // 请求成功响应回调
      onSuccess && onSuccess(res);
    } catch (err) {
      err && setError(JSON.stringify(err));
    } finally {
```

```
      setLoading(false);
    }
  };

  // 取消
  const cancel = () => {
    if (pollingIntervalTimer.current) {
      clearTimeout(pollingIntervalTimer.current);
      pollingIntervalTimer.current = null;
      status.current && (status.current = false);
    }
  };
```

使用侧提供了两个额外的配置参数，分别是 debounceInterval 和 throttleInterval，对应防抖和节流的间隔控制时间（ms），使用方式如下：

```
const { data, loading, error, request, cancel } = useRequest(
  queryCompensatoryOrderSituation,
  {
    manual: true,
    initialData: { compensatoryId,},
    debounceInterval: 1000, // 防抖
    throttleInterval: 1000, // 节流
    onSuccess: (res) => {
      console.log('success request!', res);
    },
  },
);

for(let i = 0; i < 10000; i++) {
  request();
};
```

在该 Hook 中，通过 lodash.debounce 或 lodash.throttle 来包装 request 函数主体，通过 option 中的判断来执行对应的包装体函数。

⑥ 对于涉及缓存需求的情况，可以设置一个时间窗口，在这个时间段内，如果接口没有新的请求，就直接使用上一次的响应数据。而对于依赖更新的部分，当某些状态发生改变时，需要立即发起新的请求以获取最新的数据，改造后的代码如下：

```
interface UseRequestOptionsProps {
  /** 手动开启*/
```

```typescript
  manual?: boolean;
  /** 请求参数 */
  initialData?: object;
  /** 轮询 */
  pollingInterval?: number | null;
  /** 准备，用于依赖请求 */
  ready?: boolean;
  /** 防抖 */
  debounceInterval?: number;
  /** 节流 */
  throttleInterval?: number;
  /** 延迟 loading 为 true 的时间 */
  loadingDelay?: number;
  /** 依赖 */
  refreshDeps?: any[];
  /** 请求成功回调 */
  onSuccess?: (res: any) => void;
}

const useRequest = (
  requestFn: (
    initialData?: object | string | [],
  ) => Promise<SetStateAction<any>>,
  options: UseRequestOptionsProps,
) => {
  const [data, setData] = useState<SetStateAction<any>>(null);
  const [loading, setLoading] = useState<boolean>(false);
  const [error, setError] = useState<string | null>(null);
  const status = useRef<boolean>(false);
  const pollingIntervalTimer = useRef<NodeJS.Timer | null>(null);
  const {
    manual,
    initialData,
    pollingInterval,
    ready = true,
    debounceInterval,
    throttleInterval,
    loadingDelay,
    refreshDeps,
    onSuccess,
  } = options;
```

```js
useEffect(() => {
  if (loadingDelay) {
    setTimeout(() => {
      status && setLoading(true);
    }, loadingDelay);
  }
  setError(null);
  setData(null);
  // 手动触发 request
  !manual && ready && request();
}, [manual, ready, ...(Array.isArray(refreshDeps) ? refreshDeps : [])]);

// 请求
const request = () => {
  if (debounceInterval) {
    lodash.debounce(requestDoing, debounceInterval)();
  } else if (throttleInterval) {
    lodash.throttle(requestDoing, throttleInterval)();
  } else {
    requestDoing();
  }
};
// useRequest 业务逻辑
const requestDoing = async () => {
  try {
    !status.current && (status.current = true);
    if (pollingInterval && status.current) {
      pollingIntervalTimer.current = setTimeout(() => {
        status.current && request();
      }, pollingInterval);
    }
    const res = await requestFn(initialData);
    setData(res);
    // 请求成功响应回调
    onSuccess && onSuccess(res);
  } catch (err) {
    err && setError(JSON.stringify(err));
  } finally {
    setLoading(false);
  }
};
// 取消
```

```
  const cancel = () => {
    if (pollingIntervalTimer.current) {
      clearTimeout(pollingIntervalTimer.current);
      pollingIntervalTimer.current = null;
      status.current && (status.current = false);
    }
  };
  // 缓存
  const cachedFetchData = useCallback(() => data, [data]);

  return { data, loading, error, request, cancel, cachedFetchData };
};
```

为实现缓存与依赖更新功能，我们提供了 cachedFetchData 和 refreshDeps 两个参数。在 useRequest 中获取初次数据后，我们将数据作为依赖存储于 useCallback 内，并对外公开 cachedFetchData 方法，用于平滑过渡数据状态，即从无数据（null）至获取到接口返回的数据。

依赖更新的实现思路则是在页面中将 useRequest 的一系列依赖状态一并加入 Hook 请求的副作用中，监听到页面中依赖发生变化时则重新请求，具体就是通过 refreshDeps 参数来实现，代码如下：

```
const [mountLoading, setMountLoading] = useState<boolean>(false);
const [updateLoading, setUpdateLoading] = useState<boolean>(false);

setTimeout(() => {
  setMountLoading(true);
}, 1000);

setTimeout(() => {
  setUpdateLoading(true);
}, 2000);

const { data, loading, error, request, cancel, cachedFetchData } = useRequest(
    queryCompensatoryOrderSituation,
    {
      manual: true,
      initialData: { compensatoryId,},
      debounceInterval: 1000, // 防抖
      throttleInterval: 1000, // 节流
      refreshDeps: [mountLoading, updateLoading],
      onSuccess: (res) => {
        console.log('success request!', res);
```

```
      },
    },
  );
```

至此，一个简易版本的 useRequest 实现了。其实现过程应该可以让你更深入地理解自定义 Hook 的原理和方法，并帮助你在实际业务中提高效率，从而专注于处理具体的业务逻辑。

4.11 ahooks 入门

ahooks 是一套由阿里巴巴开源的高质量高可靠的 React Hooks 库。ahooks 提供了大量开箱即用的基础 Hook 以及提炼自业务的高级 Hook；支持服务器端渲染（SSR）；对输入输出函数做了特殊处理，可有效避免闭包问题；使用 TypeScript 构建，提供了完整的类型定义文件。

在项目中使用 ahooks 非常简单，只需要安装和引入两步即可使用，就像这样：

```
安装：npm install --save ahooks
使用：import { useRequest } from 'ahooks';
```

接下来列举 ahooks 中几个常用的 Hook。

① useBoolean 是一个优雅的用于管理 boolean 类型状态的 Hook，基础用法如下：

```
import React from 'react';
import { useBoolean } from 'ahooks';

export default () => {
  const [state, { toggle, setTrue, setFalse }] = useBoolean(true);

  return (
    <div>
      <p>Effects: {JSON.stringify(state)}</p>
      <p>
        <button type="button" onClick={toggle}>
          Toggle
        </button>
        <button type="button" onClick={setFalse} style={{ margin: '0 16px' }}>
          Set false
        </button>
        <button type="button" onClick={setTrue}>
          Set true
```

```
      </button>
    </p>
  </div>
 );
};
```

  useBoolean 向外暴露了 state 和 actions 的操作集合，在操作集合中你可以直接通过 API 的形式来直接管理状态，这种方式增加了代码的可读性。如果不使用 useBoolean，更改一个状态时可能需要这样：

```
const [visible, setVisible] = useState(false);
setVisible(!visible);
```

  ② useMount 可实现只在组件初始化时被执行，也就是说在组件整个生命周期只会被执行一次，使用方式如下：

```
import { useMount } from 'ahooks';
const MyComponent = () => {
  useMount(() => {
    console.log('mount');
  });
  return <></>;
};
```

  如果不使用 useMount，可能需要像这样去写：

```
import { useEffect } from 'react';
const MyComponent = () => {
  useEffect(() => {
    console.log('mount');
  }, []);
  return <></>;
};
```

  这样看似乎并没有太大区别，但当 useUpdateEffect 出现后，开发者就会发现原生 Hook 的短板与 ahooks 的便捷了。useUpdateEffect 用法等同于 useEffect，但是会忽略首次执行，只在依赖更新时执行。如果你了解过 Vue.js，会发现这和 wacth 监听器的概念很像，监听器是比较常见的需求，但是在 React 中并没有内置该能力。useEffect 与 useUpdateEffect 的对比如下：

```
import React, { useEffect, useState } from 'react';
import { useUpdateEffect } from 'ahooks';

export default () => {
```

```
  const [count, setCount] = useState(0);
  const [effectCount, setEffectCount] = useState(0);
  const [updateEffectCount, setUpdateEffectCount] = useState(0);

  useEffect(() => {
    setEffectCount((c) => c + 1);
  }, [count]);

  useUpdateEffect(() => {
    setUpdateEffectCount((c) => c + 1);
    return () => {
      // ...
    };
  }, [count]);

  return (
    <div>
      <p>effectCount: {effectCount}</p>
      <p>updateEffectCount: {updateEffectCount}</p>
      <p>
        <button type="button" onClick={() => setCount((c) => c + 1)}>
          reRender
        </button>
      </p>
    </div>
  );
};
```

运行上例将可以看到，当 React 首次渲染时会触发一次 count 的 useEffect 副作用，但是来自 updateEffectCount 的副作用只有在点击按钮时才会发生。

ahooks 是优秀的 React Hooks 库，也是日常中开发经常会用到的库，学习理解这些 Hook 的实现原理能使我们对 React 的运行机制、底层逻辑的理解更上一层楼。本节会挑选一些 ahooks 中常用的 Hook 进行源码剖析，如 useInterval、useUpdateEffect、useSetState。

我们先看一下 useInterval，这是一个实现定时执行某段逻辑函数的 Hook，它的实现源码如下：

```
import { useCallback, useEffect, useRef } from 'react';
import useMemoizedFn from '../useMemoizedFn';
import { isNumber } from '../utils';
```

```
const useInterval = (
  fn: () => void,
  delay?: number,
  options: { immediate?: boolean } = {}
) => {
  const timerCallback = useMemoizedFn(fn);
  const timerRef = useRef<ReturnType<typeof setInterval> | null>(null);
  const clear = useCallback(() => {
    if (timerRef.current) {
      clearInterval(timerRef.current);
    }
  }, []);

  useEffect(() => {
    if (!isNumber(delay) || delay < 0) {
      return;
    }
    if (options.immediate) {
      timerCallback();
    }
    timerRef.current = setInterval(timerCallback, delay);
    return clear;
  }, [delay, options.immediate]);

  return clear;
};
```

useInterval 一共接收了三个参数：fn 代表需要执行的函数，delay 代表时间间隔，options 包含了一个 immediate 属性表示是否需要立即执行一次函数。

在 useInterval 内部，利用 useMemoizedFn 来缓存传入的函数，确保每次渲染时都是同一个函数引用，这是一个性能优化的细节。然后，使用 useRef 来保存定时器的引用，初始化值为 null。

接着，定义了一个 clear 函数用来清除定时器。在 clear 函数内部，通过判断 timerRef.current 是否存在来决定是否清除定时器。

然后在 useEffect 中，监听 delay 和 options.immediate 的变化，如果 delay 不是数字或者小于 0，则直接返回；如果 options.immediate 为 true，则立即执行一次传入的函数。再然后使用 setInterval 来设定定时器，每隔一定时间调用一次传入的函数。最后，返回 clear 函数，以便组件卸载时清除定时器。

ahooks 的 useInterval 利用了 React Hooks 的特性来管理定时器的执行，保证了组件挂载后以一定的时间间隔执行指定函数，并且基于 useEffect 在组件卸载时清除定时器，且代码中有一些性能优化的手法，值得我们学习它的封装思想。

接下来我们再看一下 useUpdateEffect。对比于 useEffect，useUpdateEffect 只会在依赖项改变时才会执行回调函数，省略了组件渲染时的一次执行，与监听触发器的概念比较类似，它在 ahooks 中的实现源码如下：

```
import { useRef } from 'react';
import type { useEffect, useLayoutEffect } from 'react';

type EffectHookType = typeof useEffect | typeof useLayoutEffect;

export const createUpdateEffect: (hook: EffectHookType) => EffectHookType
= (hook) => (effect, deps) => {
  const isMounted = useRef(false);

  hook(() => {
    return () => {
      isMounted.current = false;
    };
  }, []);

  hook(() => {
    if (!isMounted.current) {
      isMounted.current = true;
    } else {
      return effect();
    }
  }, deps);
};
```

这是 useUpdateEffect 调用创建副作用的核心函数 createUpdateEffect，该函数接收两个参数：React 中的 useEffect 或者 useLayoutEffect 和依赖项，并返回了一个新的 Hook，比较类似于 React 底层 Hook 的二次封装，该 Hook 可以在组件挂载后更新时执行副作用操作。从上到下分析一下这段代码的逻辑：

① 导入了 React 中的 useRef 和副作用函数类型；
② 定义了类型变量 EffectHookType，表示 useEffect 或 useLayoutEffect 函数类型；
③ 导出一个名为 createUpdateEffect 的函数，接收 EffectHookType 类型的参数，并返回一个新的 Hook；
④ 函数内部，使用 useRef 创建了一个名为 isMounted 的引用对象，用来存储表示当前组件是否已经挂载过的状态；
⑤ 使用传入的 Hook（即 useEffect 或 useLayoutEffect）注册一个清理函数，

当组件卸载时将 isMounted 设置为 false；

⑥ 再次使用传入的 Hook 注册一个副作用，如果组件尚未挂载，则设置 isMounted 为 true，否则执行传入的 effect 函数，并根据依赖项数组进行更新检查。

这段代码的核心思想即通过 isMounted 引用对象来判断组件状态，确保只有在组件挂载之后才会执行副作用操作，它有很多值得我们学习的地方。由此，也可以感受到理解 React 的组件生命周期是编写自定义 Hook 的前提。

接下来我们再看 useSetState。该 Hook 是 useState 的升级版本，支持部分状态更新和函数式更新，使得状态管理更加灵活和强大。其源码如下：

```
import { useCallback, useState } from 'react';
import { isFunction } from '../utils';

export type SetState<S extends Record<string, any>> =
  <K extends keyof S>(
    state: Pick<S, K> | null | ((prevState: Readonly<S>) =>
      Pick<S, K> | S | null)
  ) => void;

const useSetState = <S extends Record<string, any>>(
  initialState: S | (() => S)
): [S, SetState<S>] => {
  const [state, setState] = useState<S>(initialState);

  const setMergeState = useCallback((patch) => {
    setState((prevState) => {
      const newState = isFunction(patch) ? patch(prevState) : patch;
      return newState ? { ...prevState, ...newState } : prevState;
    });
  }, []);

  return [state, setMergeState];
};
```

该 Hook 提供了一个类似类组件 setState 的方法。它接收一个初始状态（可以是对象或者函数）作为参数，并返回当前状态和更新状态的方法，使用方式与 useState 完全一致。此外，useSetState 还进行了一层 useCallback 的包裹，保证了返回函数的引用指针唯一。

读者若有余力可以去阅读 ahooks 仓库源码中的其他 Hook，这将有助于迅速提升 React 编码水平。

思考题

1. 什么是 React Hooks？它在什么场景中使用？可以在类组件中使用 React Hooks 吗？
2. useState 是同步的还是异步的？为什么不能直接修改状态值？这样做会有什么后果？
3. 如何确保多个 useState 更新操作按顺序执行？
4. useEffect 和 useLayoutEffect 的区别是什么？什么情况下才需要使用后者？
5. 如何避免 useEffect 中的副作用无限循环？
6. useRef 的使用场景是什么？请举例说明 useRef 的一个常见应用场景。
7. useMemo 和 useCallback 是性能优化 Hook，它是如何提升性能的？
8. useMemo 和 useCallback 在父子组件的场景中，是如何与 React.memo() 实现缓存优化的？
9. 什么时候应该用 useContext？为什么不用 props 传递数据解决呢？
10. 什么是自定义 Hook？它和普通函数有什么区别？与封装组件相比，封装一个自定义 Hook 的区别是怎样的？

第 5 章

React 状态管理方案

状态，是指数据的变化；而状态管理，则是一种用于管理应用程序数据和更新用户界面的技术。在 React 中，状态主要可以分为两种：

- 视图状态，例如一个弹窗组件的 visible、Tab 组件的选中 index、loading 加载状态等；
- 逻辑状态，例如异步请求后端返回的数据（列表、详情）、某个功能里各个状态联动的数据等。

在使用 React 的开发过程中，我们常常需要处理组件间的交互与数据共享。最初，我们可能会依赖于简单的父子组件通信机制，即通过 props 将状态从父组件传递给子组件。这种方法对于小型项目来说可能足够了，但在大型应用中，随着组件数量的增长和层级的加深，这种模式会变得难以维护。

当组件之间的关系变得错综复杂时，便需要一种更为强大的工具来帮助我们管理状态。这时，引入专业的状态管理库（如 Redux、MobX 等）就显得尤为关键。这些工具不仅能够简化我们的代码，还能提高应用程序的可预测性和可测试性，从而带来更好的用户体验。

5.1 主流状态管理方案

React 生态量很庞大，拥有众多用于状态管理的方案，而我们关注的问题可以聚焦在以下三点：

- 状态管理包如何获取和设置基本的状态，这是初始化状态管理库的第一步；
- 前端状态管理大部分由异步流组成，状态管理包如何管理异步工作流；
- 数据联动是如何处理的。

目前主流的状态管理方案如下：

- 内置 Context API（createContext+useReducer+useContext）；
- Redux；

- MobX；
- rocoil。

下面分别介绍这几种状态管理方案。

（1）Context API

Context API 并不是一个状态管理工具，而是属于 React 内置的状态管理功能，其通过使用 useContext+useReducer 便可基本实现状态管理。

Context API 通过 useContext 生成一个消费状态管理的模块，如果将 state 和 dispatch 都传入 context，就可以通过 dispatch 来修改状态，就像这样：

```
const { state, dispatch } = useContext(StateContext);
// ...
dispatch({
  type: "CHANGE",
  inputValue: e.target.value,
});
```

对于异步的处理，Context API 并没有提供任何 API，需要开发者自己进行封装，或者直接在业务逻辑中处理。关于数据间的联动，Context API 也没有提供这方面的能力，同样也需要开发者去自定义封装一些方法来实现。

Context API 的优点：

- 随调随用，不需要安装第三方依赖包，属于 React 内置的能力。
- 基本可以满足简单的小范围状态管理。
- 书写方便。

Context API 的缺点：

- Context 只支持存储单一的值，当数据量大的时候，需要创建大量的 context。
- 直接使用会产生性能问题。由于 React 组件化的概念，每一次 dispatch 对 state 某个值进行变更，即使没有使用该值，也会导致其他使用该 state 的组件都重新渲染。虽然可以通过 useMemo 来解决这个问题，但是对于开发而言有很高的成本，没有通用的解决方案或者中间件来处理。

（2）Redux

Redux 是 React 开发中较常用的状态管理库，它的整体工作流程（图 5.1）为：

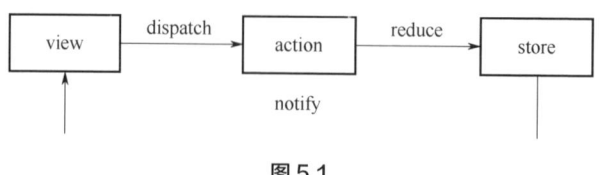

图 5.1

- 用户在view层触发某个事件，通过dispatch发送了action和payload；
- action和payload被传入reducer函数，返回了一个新的state；
- store收到reducer返回的state并进行更新，同时再去通知view层进行重新渲染。

从流程图可以看出，Redux从发起状态变更到最后响应视图更新的过程是单向数据流的模式，不可回转，因此Redux实际上也是一个纯函数，是可预测的状态变化，其遵循以下三点原则：

- 单一数据源：Redux的store只有一个，可以想象成是项目中唯一的一个大仓库，所有的状态都放在store中，所有的state共同组成了一个树形结构。
- state不可变性：在Redux中修改state的方式就是dispatch（派发）一个action，根据action的payload返回一个新的state，无法直接修改state。
- 纯函数：Redux通过reducer函数来修改状态，它每次都接收前一次的state和action，返回新的state，只要传入的state和action相同，返回的结果也一定相同，这也是纯函数的原则。

Redux没有规定如何处理异步数据流，最原始的方式就是使用action creators，也就是制造action之前进行异步操作，然后把复用的操作抽离出来。当然在实际项目中通常借助于redux-thunk、redux-saga这些中间件包来支持处理异步。

Redux的优点：

- 繁荣的社区：像支持异步变更的中间件就有很多社区的方案来支持，并且遇到的问题在社区中都可以找到答案。
- 可扩展性很高：中间件模式可以让开发者可以自己开发出符合自身业务需求的中间件。
- 单一数据源并且是树形结构，这让Redux支持回溯。

Redux的缺点：

- 大量的模板代码：在下节具体介绍Redux的使用时，会发现其写起来很累人，这是Redux对比其他方案最大的劣势。
- 状态量大了以后，会出现性能问题，因为所有的state都存往store，每一次action（行动）都要将所有reducer执行一遍，耦合比较严重。

（3）MobX

MobX是一个非常典型的响应式状态管理工具，它的工作流程大概如下：

- 用户在view触发某个事件（和Redux的action一样）；
- 事件触发action执行，通过action来修改state（和Redux的reducer一样）；
- state更新后，computed values会根据依赖重新计算属性值（和Vue的计算属性概念一样）；

- 状态更新后会触发 reactions，来响应式操作这一次状态变化（重新渲染组件）。

因此 MobX 这种响应式设计和 Vue.js 的 MVVM 设计模型很类似，与 Redux 不同，MobX 对全局 state 做了一层代理，监听 state 的变化，当 state 变化时，会自动更新相关的计算属性，所以 MobX 是直接修改 state。通过 MobX 触发一次状态管理变更的流程如图 5.2 所示。

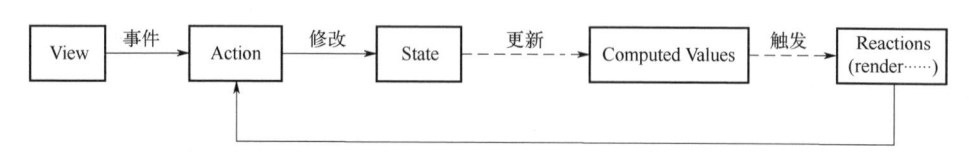

图 5.2

那么 MobX 是如何处理异步调配的呢？由于 MobX 中的 action 指的就是一段可以修改 store 中 state 的代码，就像是这样：

```
import { makeAutoObservable } from "mobx";

class TodoStore {
    todos = [];

    constructor() {
        makeAutoObservable(this);
    }

    addTodo = (todo) => {
        this.todos.push(todo);
    }

    removeTodo = (index) => {
        this.todos.splice(index, 1);
    }
}

const todoStore = new TodoStore();
export default todoStore;
```

因此，MobX 对于状态的处理是非常自然、面向对象的，很容易上手，代码风格也是比较放开的，这也是笔者日常使用最多的状态管理库。

MobX 的优点：上手简单，没有太多概念性的 API，只需了解 MobX 的架构和设计，在 store 里声明 state 和修改 state 的方法就可以直接用了。

MobX 的缺点：风格放开其实是一把双刃剑，如果没有统一团队的代码风格、技术水平参差不齐，那么会在项目中看到各种各样 MobX 的代码。

（4）recoil

recoil 是 React 官方推出的一个状态管理库，其设计的思路是将状态原子化，atom 和 selector 是 recoil 的两个核心概念。

- atom：一个原子是一个共享状态的片段。
- selector：一个组件可以订阅一个原子来获取或设置它的值。

recoil 把每一个状态都定义为一个原子，每一个原子都是可订阅、可修改的单元，跨越在整个 React 应用中，在消费一个状态时，需要 import 两个函数：

```
import { useRecoilState } from "recoil";
import { xxxState } from "../store";
useRecoilState(xxxState);
```

那么 recoil 是如何处理异步操作的呢？首先准备一个异步函数，用 selector 来包裹，定义为一个 dispatch：

```
const userInfo = selector({
  key: "userInfo",
  get: async () => {
    const res = await getUserInfo();
    return res.name;
  }
});
```

然后使用 useRecoilValueLoadable 来消费，类似 Redux 中的 action：

```
const Info = () => {
  const userInfoLoadable = useRecoilValueLoadable(userInfo);
  switch (userInfoLoadable.state) {
    case "hasValue":
    // ...
    case "loading":
    // ...
    case "hasError":
    // ...
  }
};
```

对于异步操作，recoil 无法提供天然的编码方式，但是官方给出了对应的异步 API，只是代码相对繁琐，有一些包装成本。

recoil 的优点：

- React 官方推出的状态管理方案，有保障性，不用担心库不维护这类事。
- 对于 React concurrent 模式支持良好。

recoil 的缺点：
- 不支持类组件，如果是旧项目的话可以直接忽略这个方案了，除非重构到函数组件架构。
- API 很多，上手成本高。
- 消费一个状态很繁琐，需要 import 两个官方的 Hook 函数。
- 原子化概念对比其他几种状态管理方案，比较难理解，很难恰当使用。

至此，相信你对于 Context API、Redux、MobX、recoil 这四种状态管理解决方案都有了些概念性的了解。这几个方案各有所长，需要按照实际环境来进行技术选型，我们可以根据它们的优缺点来判断实际业务中采用哪种方案：

- 如果开发的应用中，每个页面各个部分功能都比较独立，没有什么关联性，那么可以直接用父组件 props 传子组件就可以满足需求。
- 如果组件跨度远，但也是模块性的，并且状态管理逻辑不是很复杂，那么 Context API 也足以满足需求。
- Redux、MobX、recoil 各有所长，如果项目组对于 Redux 有一些封装和规范，那么可以选择 Redux。
- 如果共享状态不算多，觉得 Redux 太复杂，则可以考虑 MobX 或 recoil。
- 如果对于 MobX 有一些封装和规范化，那么请直接使用 MobX。

5.2 Redux

本节主要介绍 Redux 的使用。首先，让我们来细致了解下的 Redux 具体工程流程。

如图 5.3 所示，Redux 共由三个核心部分组成，分别是 action、reducer、store。在 React 生态圈中大部分状态管理都是基于这三者构成的，只是设计思想上会有不同侧重。

action 代表了动作的对象，表示对状态库进行的变更操作。例如，对状态库中的 num 进行 feat 操作，就会使 num+1。action 包含了两个属性：

- type:属性标识，值为字符串，是必要属性，对应操作动作（feat/fix/refactor）。
- data:数据属性，值为任意类型，是可选属性，即代表变更状态填入的指定值。

在 action 变更加工数据时，会根据旧的 state 和 action，产生出新的 state，因此只要 type、data 相同，每次变更的结果就相同，因此 Redux 的变更也是一个很经典的纯函数。

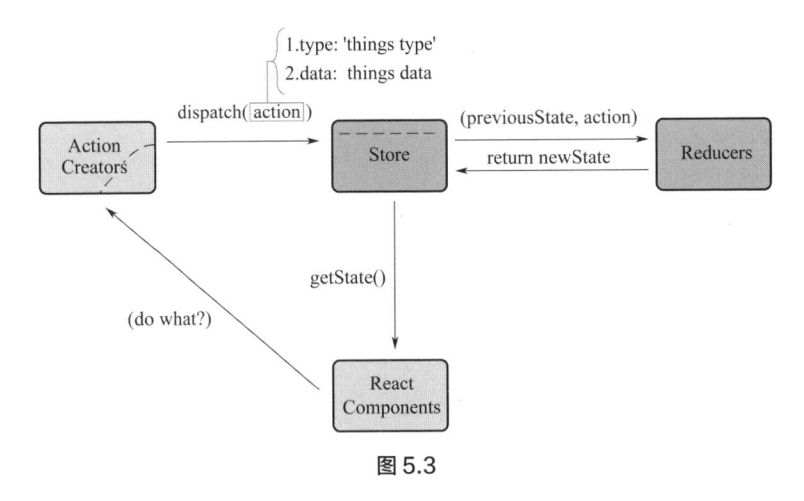

图 5.3

reducer 用于初始化、更新状态。在加工的过程中，其会根据不同的 state 和 action 产生出新的 state。

store 是将 state、action、reducer 联系在一起的对象，加工时会根据旧的 state 和 action，产生出新的 state 的纯函数。

下面，我们开始 Redux 的应用实践。

① 在项目中安装 Redux 包：

```
npm i redux
```

② 搭建出这样的一个单独的 store 目录结果，用于存放所有与 Redux 相关的代码：

```
└─ store
   ├─ actions //actions，文件夹内以模块区分
   │  ├─ count.js
   │  └─ person.js
   ├─ constants.js //action type 唯一标识常量
   ├─ index.js //入口文件
   ├─ reducers //reducers，文件夹内以模块区分
   │  ├─ conut.js
   │  └─ index.js //reducers 统一暴露文件，合并 reducers
   └─ persons.js
```

在使用 Redux 的过程中，应遵循模块化的规范，将不同的功能模块拆解到不同的文件中，主要应用于 action 和 reducer 中。

③ 引入 createStore，专门用于创建 Redux 中最为核心的 store 对象，而

redux-thunk、applyMiddleware 用于支持异步 action（后面会讲到什么是异步 action）：

```
//src/store/index.js
import { createStore, applyMiddleware } from "redux";
//用于支持异步 action
import thunk from "redux-thunk";
import reducers from "./reducers";
export default createStore(reducers, applyMiddleware(thunk));
```

④ 定义 action 对象中 type 类型的变量枚举值，消除魔法字符串，增加开发规范，代码如下：

```
//src/store/constants.js
export const INCREMENT = 'increment'
export const DECREMENT = 'decrement'
```

⑤ 开始创建 action，分为同步 action 和异步 action，入参就是 type 和 data，代码如下：

```
// src/store/actions/count.js
import { INCREMENT, DECREMENT } from "../constants";
// 普通 action 的值为 object `{type: INCREMENT, data }`
export const increment = data => ({ type: INCREMENT, data });
export const decrement = data => ({ type: DECREMENT, data });
// 异步 action 的值为函数
export const incrementAsync = (data, time) => {
  return (dispatch) => {
    setTimeout(() => {
      dispatch(increment(data));
    }, time);
  };
};
```

注释标注的异步 action，就是异步事件中涉及的对于 store 的变更操作，但是延迟的操作不能交给组件本身，而应该交给 action，否则会出现数据延迟同步的情况。那么何时需要采用异步 action 呢？当想要对状态进行操作，但是具体的数据依赖异步任务结束时，就需要采用异步 action。

⑥ 从 reducer 函数中会连接到两个参数，分别是之前的状态（preState）和动作对象（action）。我们可以通过 preState 和 action 中的 type 来决定此次数据如何加工，代码如下：

```
import {INCREMENT, DECREMENT} from '../constants'
// 初始化状态
```

```
const initState = 0;

export default function count(preState = initState, action) {
  const { type, data } = action;
  switch (type) {
    case INCREMENT:
      return preState + data;
    case DECREMENT:
      return preState - data;
    default:
      return preState;
  }
}
```

这样我们就可以在组件中使用 Redux 来测试状态的变更了。对于 reducer，有几个需要特殊注意的点：

第一点，在 reducer 里面是没有默认初始值的，因此在控制台中打印会发现 preState=undefined，type=@@redux/INITq.p.v.o.d.w，type 为随机值是因为在 Redux 内部处理时，每次的输出都会不一样，如图 5.4 所示。

图 5.4

那么如果要给 reducer 设置默认初始值，可以通过以下代码设置：

```
const initState = 1;
function count(preState = initState) {}
```

第二点，reducer 是一个纯函数，因为在 reducer 中，我们有了明确的 switch 分支判断，每一次 type 都有精准的链路，可以确保有相同的入参，不会被外部作用域干扰，每次的返回值都是相同的。

纯函数的概念如下：

- 一类特别的函数，只要输入是相同的，必定得到相同的输出；
- 必须遵循一些约束：不得二次改写参数数据；不会产生任何副作用，例如网络请求；不能调用 Date.now() 或者 Math.random() 等任何不纯、带有随机性风险的函数。

通过 combineReducers 可以进行 reducer 合并，最后集成在 store 入口文件中：

```
// src/store/reducers/index.js
import { combineReducers } from "redux";
import count from "./conut";
import persons from "./persons";

export default combineReducers({
  count,
  persons,
});
```

最后到了应用层，我们如何在组件中消费 Redux？那可太简单了！就像这样：

```
import store from "../../store";
import { increment } from "../../store/action/count";

// 加法
clickIncrement = () => {
  const { value } = this.selectNumber;
  store.dispatch(increment(value * 1));
};
  render() {
    return (
      <div>
        <h1>当前求和为：{store.getState()}</h1>
        ...
        <button onClick={this.clickIncrement}>+</button>
      </div>
    )
  }
```

在组件中，我们只需要通过 getState() 获取 store 的数据，通过 dispatch 触发 store 的更新即可，所有的复杂工程代码都在 store 文件中。

5.3 react-redux

为了方便在 React 中进行状态管理，Redux 的作者封装了一个 React 专用的状态管理库——react-redux，它提供了一种更高效的方式来操作 Redux 状态。那它和 Redux 有什么区别呢？

- Redux 需要利用 store.subscribe 监听 store 的变化来更新视图，react-redux 则不需要这个操作。
- 设计模式不同：react-redux 将组件分为了 UI 组件、容器组件，关于状态管理的操作都在容器组件中，职责明确。它通过 connect 将容器组件和 UI 组件链接。

react-redux 将所有的组件分为了两大类：UI 组件（presentational component）和容器组件（container component）。

UI 组件的特性主要如下：
- 只负责 UI 的呈现，不带有业务逻辑；
- 无状态；
- 所有数据由 props 提供；
- 不使用 Redux。

容器组件的特性正好相反：
- 负责管理数据和业务逻辑，不负责 UI 展示；
- 有状态；
- 使用 Redux。

那 UI 组件与容器组件如何关联呢？首先，所有的 UI 组件都应该包裹一个容器组件，它们是父子关系，容器组件与 Redux 通信，并给下游（UI 组件）提供数据，通过 props 传递数据。

总结一下，UI 组件负责 UI 的呈现，容器组件负责管理数据和逻辑，如果一个组件既有 UI 又有业务逻辑，则应该拆为父子结构：外层是一个容器组件，里面包含了一个或多个 UI 组件，前者负责与外部的通信，将数据传递给后者，后者负责渲染视图。

react-redux 提供了 connect 方法，用于从 UI 组件生成容器组图。在下面代码示例中，CountUI 是 UI 组件，利用 connect 方法导出了一个容器组件，为了定义业务逻辑，需要给出下面两个信息：
- 输入逻辑：外部数据（state）如何转换为 UI 组件的参数；
- 输出逻辑：用户发出的动作如何变为 action 对象，从 UI 组件传出去。

connect 方法接收这两个参数，对应：mapStateToProps 和 mapDispatchToProps。它们定义了 UI 组件的业务逻辑。前者负责输入逻辑，即将 state 映射到 UI 组件的参数（props），后者负责输出逻辑，即将用户对 UI 组件的操作映射成 action 交给 react-redux。mapStateToProps 接收 state 参数，mapDispatchToProps 接收 dispatch 参数，代码如下：

```
// 容器组件
import { connect } from "react-redux";
import CountUI from "../../components/count";
import {
createIncrementAction,
```

第 5 章　React 状态管理方案　175

```
  createDecrementAction,
  createIncrementAsyncAction,
} from "../../redux/count_action";

const mapStateToProps = (state) => ({ count: state });

const mapDispatchToProps = (dispatch) => ({
  increment: (number) => {
    dispatch(createIncrementAction(number));
  },
  incrementAsync: (number) => {
    dispatch(createIncrementAsyncAction(number, 500));
  },
  decrement: (number) => {
    dispatch(createDecrementAction(number));
  },
});

export default connect(mapStateToProps, mapDispatchToProps)(CountUI);
```

UI 组件代码如下：

```
// UI 组件
import React, { Component } from "react";

export default class CountUI extends Component {
  // 加法
  increment = () => {
    const { value } = this.selectNumber;
    this.props.increment(value * 1);
  };
  // 减法
  decrement = () => {
    const { value } = this.selectNumber;
    this.props.decrement(value * 1);
  };
  // 奇数加
  incrementIfOdd = () => {
    if (this.props.count % 2 === 1) {
      const { value } = this.selectNumber;
      this.props.increment(value * 1);
    }
  };
```

```
// 异步加
incrementAsync = () => {
  const { value } = this.selectNumber;
  this.props.increment(value * 1);
};
render() {
  return (
    <div>
      <h1>当前求和为：{this.props.count}</h1>
    </div>
  );
}
}
```

　　mapDispatchToProps 是 connect 函数的第二个参数，用于建立 UI 组件的参数到 store.dispatch 方法的映射，换言之，它定义了哪些用户的操作应该作为 action 传给 store。mapDispatchToProps 可以是一个函数（代码块例子），也可以是一个对象。如果是函数，在 UI 组件中则为自动执行这个函数，将对象返回出去；如果是一个对象，直接消费即可，返回的 action 会由 Redux 自动发出，就像这样：

```
// mapDispatchToProps 的简写，返回 object
export default connect(mapStateToProps, {
  increment: createIncrementAction,
  incrementAsync: createIncrementAsyncAction,
  decrement: createDecrementAction,
})(CountUI);
```

　　如果像上面这样写，看起来是美好的，但如果容器组件在很深的层级，一级一级传下去就会很麻烦，该如何解决呢？react-redux 提供的 Provider 组件，可以让容器组件得到 state，这也是一种透传的思想，只需要让 Provider 在根组件外面包一层，App 中所有子组件都可以默认得到 state 了。原理就是 React 中的 context 上下文属性！使得原来的整个应用成为 Provicer 的子组件，接收 Redux 的 store 作为 props，通过 context 对象传递给下级组件上的 connect 即可，就像这样：

```
// src/index.js
import React from "react";
import ReactDOM from "react-dom";
import { Provider } from "react-redux";
import store from './redux/store'
import App from "./App";
ReactDOM.render(
  <Provider store={store}>
```

```
    <App />
  </Provider>,
  document.getElementById("root")
);
```

可见，react-redux 的使用显得更轻量简洁。

5.4　实现一个简易版 Redux

本节我们将动手实现一个简易版的 Redux，以便于深入理解它的工作原理。

先回顾一下 Redux 的基本使用。如下例所示，包含了 Redux 中的 store、action 和 reducer 三大概念：

```
import { createStore } from 'redux';

const initState = {
  milk: 0
};
function reducer(state = initState, action) {
  switch (action.type) {
    case 'PUT_MILK':
      return {...state, milk: state.milk + action.count};
    case 'TAKE_MILK':
      return {...state, milk: state.milk - action.count};
    default:
      return state;
  }
}

let store = createStore(reducer);

// subscribe 的本质是订阅 store 的变化
// 若 store 发生变化，传入的回调函数就会被调用
// 如果是结合页面更新，更新的操作就是在这里执行
store.subscribe(() => console.log(store.getState()));

// 将 action 发出去要用 dispatch
store.dispatch({ type: 'PUT_MILK' }); // milk: 1
store.dispatch({ type: 'PUT_MILK' }); // milk: 2
store.dispatch({ type: 'TAKE_MILK' }); // milk: 1
```

要动手实现并替换 Redux 库，就需要知道它都有些什么东西。上述代码只体

现了一个 createStore API，它接收 reducer 函数作为参数，返回一个 store，而所有的功能都在 store 上。

store 中包含以下能力：

- subscribe：订阅 state 变化，用于触发页面更新；
- dispatch：发出 action 的方法，每次 dispatch action 都会执行 reducer 生成新的 state，然后去执行 subscribe；
- getState：简单方法，返回当前的 state。

接下来我们开始动手实现 createStore：

```
const createStore = (reducer: any) => {
  let state: any;
  let listeners: Array<() => void> = [];

  /**
   * @description: 订阅，添加订阅器
   * @param {function} fn
   */
  const subscribe = (fn: () => void) => {
    listeners.push(fn);
  }

  /**
   * @description: 执行变更，触发所有回调
   * @param {*} action
   */
  const dispatch = (action: any) => {
    state = reducer(state, action);
    listeners.forEach(fn => fn());
  }

  /**
   * @description: 获取当前 state
   * @return {*}
   */
  const getState = () => {
    return state;
  }

  return {
    subscribe,
    dispatch,
```

```
    getState
  }
}

export default createStore;
```

使用我们所实现的简易版 Redux，再次执行本节的首例代码，结果如图 5.5 所示。和预期效果一致。

```
PUT_MILK
dispatching { type: 'PUT_MILK', payload: { count: 1 } }
{ milk: 1 }
next state { milk: 1 }
PUT_MILK
dispatching { type: 'PUT_MILK', payload: { count: 1 } }
{ milk: 2 }
next state { milk: 2 }
TAKE_MILK
dispatching { type: 'TAKE_MILK', payload: { count: 1 } }
{ milk: 1 }
next state { milk: 1 }
```

图 5.5

接下来我们继续完善它。当 store 庞大的时候，以及有多个模块化 reducer 的时候，我们需要使用 combineReducers API 来合并 reducer 再消费，就像这样：

```
const reducer = combineReducers({milk: milkReducer, rice: riceReducer})
```

让我们来实现这个 API。首先分析入参，就是输入多个 reducer，输出一个 reducer：

```
const combineReducers = (reducerMap: { [key in string]: any }) => {
  const reducerKeys = Object.keys(reducerMap); // 得到所有 reducer map
  const reducer = (state = {}, action) => {
    const newState = {};
    for (let i = 0; i < reducerKeys.length; i++) {
      // reducerMap 里面每个键的值都是一个 reducer，我们把它取出来运行下就可以得
      到对应键新的 state 值
      // 然后将所有 reducer 返回的 state 按照参数里面的 key 组装好
      // 最后再返回组装好的 newState 就行
      const key = reducerKeys[i];
      const currentReducer = reducerMap[key];
      const prevState = state[key];
      newState[key] = currentReducer(prevState, action);
    }

    return newState;
  };
  return reducer;
```

};

我们改一下测试用例：

```
// 使用 combineReducers 组合两个 reducer
const reducer = combineReducers({milkState: milkReducer, riceState: riceReducer});

let store = createStore(reducer);

store.subscribe(() => console.log(store.getState()));

// 操作
store.dispatch({ type: 'PUT_MILK', count: 1 }); // milk: 1
store.dispatch({ type: 'PUT_MILK', count: 1 }); // milk: 2
store.dispatch({ type: 'TAKE_MILK', count: 1 }); // milk: 1

// 操作 rice 的 action
store.dispatch({ type: 'PUT_RICE', count: 1 }); // rice: 1
store.dispatch({ type: 'PUT_RICE', count: 1 }); // rice: 2
store.dispatch({ type: 'TAKE_RICE', count: 1 }); // rice: 1
```

运行一下，结果如图 5.6 所示。官方的实现方式也是这样的，只是在中间加入了更多的异常处理。

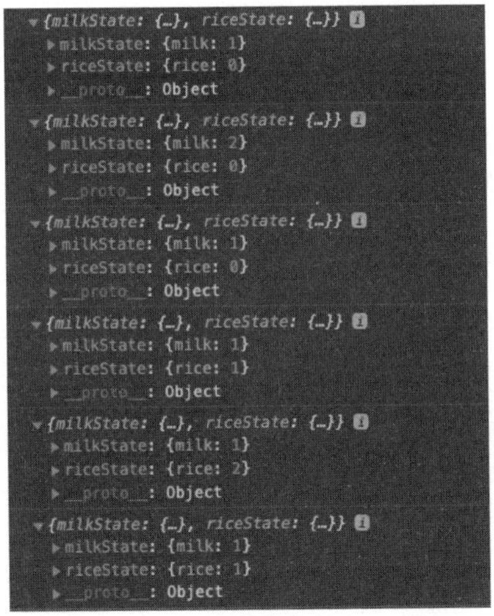

图 5.6

最后，我们来实现 applyMiddleware，也就是 Redux 中间件。我们可以在状态变更中插入变更日志，就像这样：

第 5 章　React 状态管理方案　181

```js
function logger(store) {
  return function(next) {
    return function(action) {
      console.group(action.type);
      console.info('dispatching', action);
      let result = next(action);
      console.log('next state', store.getState());
      console.groupEnd();
      return result
    }
  }
}

// 在 createStore 的时候将 applyMiddleware 作为第二个参数传进去
const store = createStore(
reducer,
applyMiddleware(logger)
)
```

可以看到,中间件在 createStore 中作为第二个参数传入,官方称之为 enhancer,即增强器,用于增强辅助 store 的能力。对此,在动手实现时,就需要改造一开始的 createStore 方法了,就像这样:

```js
const createStore = (reducer, enhancer) = () => {}
```

并加入判断:

```js
const createStore = (reducer, enhancer) = () => {
  if(enhancer && typeof enhancer === 'function'){
    const newCreateStore = enhancer(createStore);
    const newStore = newCreateStore(reducer);
    return newStore;
  }
  // 执行之前的逻辑
}
```

如果传入了 enhancer 并且是个函数,执行新的逻辑,否则就执行之前实现的逻辑。

结合之前 logger 中间件的结构可以知道 applyMiddleware 包含了两层函数,因此我们也把 applyMiddleware 的结构先设计出来。

```js
function applyMiddleware(middleware) {
  function enhancer(createStore) {
```

```
    function newCreateStore(reducer) {
      // ...
    }
  }
}
```

那这里面做了什么呢？中间件本质是加强 dispatch，我们得到加强后的 dispatch，最后再把 dispatch 和原始 store 返回即可。

```
function applyMiddleware(middleware) {
  function enhancer(createStore) {
    function newCreateStore(reducer) {
      const store = createStore(reducer);
      // dispatch 加强函数
      const func = middleware(store);
      // 解构出原始的 dispatch
      const { dispatch } = store;
      // 得到增强版的 dispatch
      const newDispatch = func(dispatch);
      return {...store, dispatch: newDispatch}
    }
    return newCreateStore;
  }
  return enhancer;
}
```

至此，再运行一下测试用例，得到了对应 logger 中间件的结构，如图 5.7 所示。

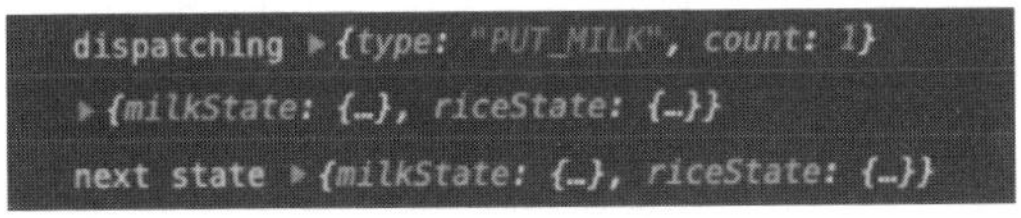

图 5.7

那如何实现多个中间件呢？当有多个中间件时，入参的数量就可能是无限的了，因此我们需要一个组合函数，把所有函数串联起来：

```
const compose = (f1, f2, f3, f4) => {
  return (args) => f1(f2(f3(f4(args))))
}
```

在 middleware 中，将多个中间件不断地升级 dispatch，就像这样：

```
function applyMiddleware(...middlewares) {
  function enhancer(createStore) {
    function newCreateStore(reducer) {
```

```
    const store = createStore(reducer);
    // dispatch 加强函数
    const chain = middlewares.map((middleware) => middleware(store));
    // 解构出原始的 dispatch
    const { dispatch } = store;
    // 用 compose 得到一个组合了所有 newDispatch 的函数
    const newDispatchGen = compose(...chain);
    // 获得加强版 dispatch
    const newDispatch = newDispatchGen(dispatch);
    return { ...store, dispatch: newDispatch };
    }
    return newCreateStore;
  }
  return enhancer;
}
```

最后再加个 logger 来测试下：

```
function logger2(store) {
  return function(next) {
    return function(action) {
      let result = next(action);
      console.log('logger2');
      return result
    }
  }
}

let store = createStore(reducer, applyMiddleware(logger, logger2));
```

新的 logger 也打印出来了。至此，一个简易版的 Redux 完成。

思考题

1. 什么是状态管理？和 useState 有什么区别？
2. 主流的状态管理解决方案有哪些？各自的特征是什么？
3. 请描述 Redux 的基本架构和工作原理。
4. Redux 是如何处理异步 action 的？
5. react-redux 库和 Redux 有什么区别？前者是如何桥接 React 和 Redux 进行协同工作的？
6. 实现一个简易版 Redux 的整体思路是怎样的？过程中用了哪些设计模式？

第 6 章

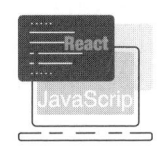

全栈化与 Serverless

随着互联网行业的饱和，全栈化将慢慢成为主流，无论是过去还是现在，只要你掌握 HTML、CSS、JavaScript、Vue/React 的技术，就能找到一份收入尚可的前端开发工作。并且现在的开发模式仍然是前后端分离的开发模式，开发一个项目的时候，前后端是最基本的研发配置，前端工程师负责视觉页面交互的开发工作，后端工程师负责数据服务的开发工作。但如果开发者同时掌握前后端的技能，就可以一个人包揽项目的整个研发资源，这就是全栈化的必要性。

前端工程师向 Nest.js、Midway.js 全栈化方向发展的原因可以归纳为以下几点：

- JavaScript/TypeScript 的通用性：Nest.js 基于 Node.js，并且可以使用 TypeScript 进行开发，这与前端开发者熟悉的技术栈非常吻合，也就意味着前端工程师可以使用已经掌握的语言，在进行一定后端概念的学习和积累后，无缝切换到后端开发。

- 全栈背景的需求增长：在当前的技术市场中，能够同时处理前端和后端的全栈工程师越来越受欢迎。这种一站式的技能集合可以使开发过程更加高效，降低沟通成本，同时也能为个人职业发展创造更多的机会。

- Nest.js 的优良设计：Nest.js 是一个高度模块化的，用于构建高效、可靠和可扩展服务器端应用程序的框架。它借鉴了 Angular 的设计模式，让熟悉 Angular 的前端工程师更加容易过渡到 Nest.js 的后端开发上。

- 扩展个人技术栈的趋势：随着开发复杂性的增加和项目需求的多样性，前端工程师往往需要了解更多技术层面的知识，不能只关注浏览器端。学习 Nest.js 能帮助前端工程师更好地理解后端逻辑、数据库交互等，从而完成更加复杂的应用开发。

- 开发体验和生态系统：Nest.js 提供了良好的开发体验和健全的生态系统，它有清晰的文档、活跃的社区支持以及丰富的第三方库，所有这些都为前端工程师提供了便利的条件来扩展其后端开发技能。

Nest、Midway 作为现代化的后端框架，为前端工程师向全栈工程师转变提供了相当便利的条件，并且这样的转变也很符合市场趋势和个人职业发展的需要。

本章先学习 Nest.js，它是目前互联网环境下，前端人员接触服务端最适合的框架，因为它基于 JavaScript 语言实现，是面向未来发展的后端框架；学习完 Nest.js 后再学习 Midway.js，该框架背靠阿里，有着国内上千产品的实战化成果，历经"双11""双12"等高并发的业务场景的考验，可靠度较高。并且在学习过程中会穿插讲解实际项目场景案例从而实现最佳的场景化学习。这两款框架是最好的全栈框架，掌握它们对前端工程师的发展意义非凡。

6.1 Nest.js 快速入门

Nest.js 是一个高度可扩展、面向服务器端的框架，用于构建有效且可靠的 Node.js 服务端应用程序。它使用 TypeScript 进行开发（也允许开发者使用纯 JavaScript）并且结合了面向对象编程、函数式编程和响应式编程的领先概念。

以下是 Nest.js 的一些特点和优势：

- 架构：Nest.js 提供了出色的模块化架构，它允许开发者通过使用模块、控制器和服务来组织代码；
- 支持强类型：JavaScript 是弱类型语言，但是 Nest.js 构建在 TypeScript 之上，这意味着它支持静态类型检查和最新的 JavaScript 特性；
- 灵感来源：Nest.js 受到了 Angular 的启发，并使用了类似的装饰器和依赖注入机制，这使得来自 Angular 背景的开发者在使用 Nest.js 时会感到非常熟悉；
- 依赖注入：Nest.js 利用 TypeScript 的特性和装饰器来提供一个全面的依赖注入系统，类似于 Java 的 SpringBoot；
- CLI 工具集成：Nest.js 提供了 CLI 工具，用于自动化开发任务，如项目初始化、生成模块、控制器和服务等，从而提高开发效率；
- 微服务支持：Nest.js 内置微服务支持能力，支持多种微服务传输层如 tcp、redis 等；
- GraphQL 和 REST API：Nest.js 使得创建 GraphQL 和 REST API 变得易如反掌，这对于构建现代 Web 应用程序来说是非常重要的；
- 丰富的生态系统：Nest.js 拥有丰富的模块和库，可以用来集成各种功能，如数据库 ORM、身份验证、CQRS 等；
- 灵活性：尽管 Nest.js 为应用程序架构提供了指导，但它仍然保持了足够的灵活度，允许开发者按照自己的要求来定制和扩展框架。

开发者通常会选择 Nest.js 来构建易维护、易扩展的大型应用程序，因为它的结构皆在支持复杂的场景和业务逻辑。Nest.js 是当今市面上构建企业级 Node.js 应用程序的一个非常流行的选择。

快速创建一个 Nest 项目非常简单，全局安装脚手架并启用严格模式创建即可，就像这样：

```
# 全局安装脚手架
npm i -g @nestjs/cli
# 启用 Typescript 严格模式创建项目
nest new project01 --strict
```

创建好新的 Nest.js 项目后，让我们来查看其项目中的关键文件，如图 6.1 所示。

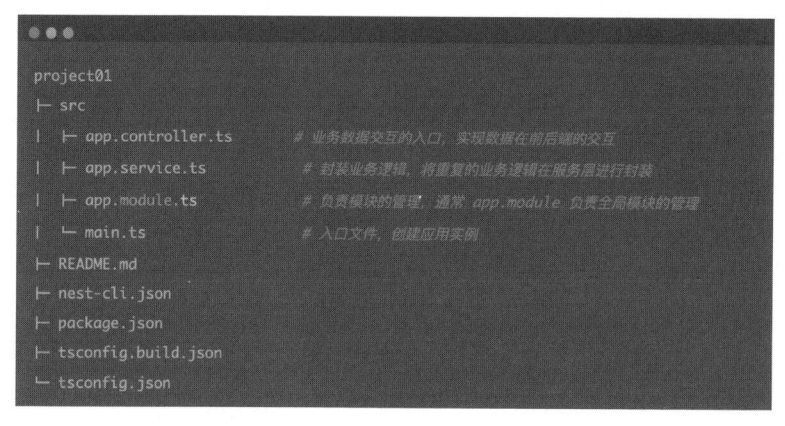

```
project01
├── src
│   ├── app.controller.ts      # 业务数据交互的入口，实现数据在前后端的交互
│   ├── app.service.ts         # 封装业务逻辑，将重复的业务逻辑在服务层进行封装
│   ├── app.module.ts          # 负责模块的管理，通常 app.module 负责全局模块的管理
│   └── main.ts                # 入口文件，创建应用实例
├── README.md
├── nest-cli.json
├── package.json
├── tsconfig.build.json
└── tsconfig.json
```

图 6.1

和常见前端项目一样，src 目录是主要的源码目录，主要由入口文件 main.ts 和一组 module、service、controller 组成。module、service、controller 通常也被称为 user 三件套，module 模块负责模块的管理，通常 app.module 负责全局模块；controller 代表业务数据交互的入口，也是暴露给前端接口的地址，通常是一个个函数；service 代表封装业务逻辑，重复的业务逻辑在 service 层封装即可。

而如果需要新增一个用户（user）模块，则新增这样的 user 三件套即可。

接下来我们需要让后端服务在本地运行起来，运行应用程序，主要分为三种启动的方式，如下：

- 普通启动模式：npm run start
- 监听启动模式：npm run start:dev
- 调试启动模式：npm run start:debug

接下来我们从模块管理开始讲解。Nest.js 是经典的采用模块化组织应用程序的框架，整个应用由一个根模块（Application Module）和多个功能模块共同组成，如图 6.2 所示。

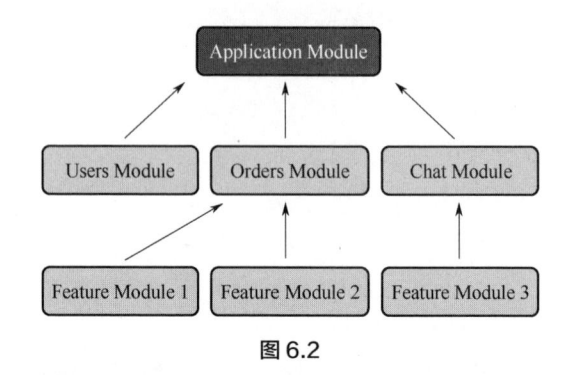

图 6.2

从图 6.2 可以看到，这是一个很明显的模块化架构方案，它其实和 Webpack 打包的样子很相似。那我们如何创建一个新的模块呢？方式很简单，Nest.js 提供了内置的 CLI 命令，在项目根目录执行以下命令即可：

- 完整命令：nest generate module <module-name>
- 简写命令：nest g mo <module-name>

例如我们生成了一个 order 服务，也就是 order 模块，生成的模块代码就像这样：

```typescript
//order.module.ts
import { Module } from '@nestjs/common';
@Module({
providers: [],
imports: [],
controllers: [],
exports: [],
})
export class OrdersModule {}
```

每个模块都是一个由@Module()装饰器注解的类，应用中模块之间的关系由@Module()装饰器中携带的所有元数据描述来产生关联。这四个元数据的能力介绍是这样的：

Provieders：注册订单提供者模块，如负责订单 CRUD 的服务；
Controllers：注册订单控制器模块，如负责订单 CRUD 的接口路由处理；
Imports：注册与订单相关联的模块，如与订单关联的用户查询服务；
Exports：导出订单提供者模块，如用户查询订单提供者统计订单数量。

通过 exports 将模块导出的行为被称为模块共享。Nest.js 中有一种概念叫模块再导出。一个模块仅负责将一系列相关联的模块通过 imports 导入，紧接着就通过 exports 全部导出的行为就是模块再导出。利用这个能力，可以减少大量关联模块重复导入造成的负担，就像这样：

```
@Module({
  imports: [DatabaseModule, RedisModule, MongoModule],
  exports: [DatabaseModule, RedisModule, MongoModule],
})
export class ConnectionModule {}
```

如果希望某个模块可以成为全局模块，在任何地方都可以开箱即用，那可以为其增加@Global()装饰器，就像这样：

```
@Global()
@Module({
  imports: [DatabaseModule, RedisModule, MongoModule],
  exports: [DatabaseModule, RedisModule, MongoModule],
})
export class ConnectionModule {}
```

接下来学习控制器的使用。先看一张图来了解一下控制器的概念以及其在前后端的主要作用，如图 6.3 所示。

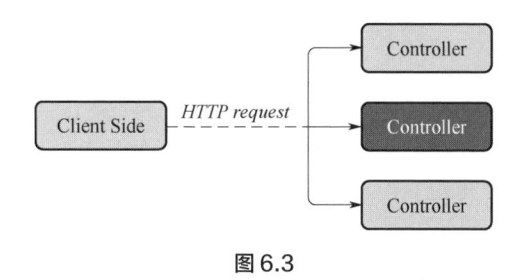

图 6.3

控制器用来接收和处理客户端发起的特定请求，不同的客户端请求将由 Nest.js 路由机制分配到对应的控制器进行处理。Nest.js 中的路由、控制器分配机制就和前端的路由、页面机制一样。

在项目中创建控制器非常简单，Nest.js 同样提供了内置的 CLI 方案，分为完整命令和简写命令，如下：

- 完整命令：nest generate controller <controller-name>
- 简写命令：nest g co <controller-name>

控制器是使用@Controller("path")来装饰注解的类，其中 path 是一个可选的路由路径前缀，通过 path 可以将相关的路由进行分组，就像这样：

第 6 章 全栈化与 Serverless 189

```
import { Controller, Get } from '@nestjs/common';
@Controller('orders')
export class OrdersController {
  @Get()
  index() {
    return 'This is the order controller';
  }
}
```

当客户端通过 GET 方案请求，对 orders 路由发送请求时将有 index()处理函数响应，请求路径就像是这样的：http://localhost:3000/orders/index。除了@Get()装饰器以外，Nest 为 HTTP 标准方法提供的请求事件装饰器还有@Post()、@Put()、@Delete()、@Patch()、@Options()和@Head()，以及@All()用来处理所有的情况。@Controller("path")中的路径从设计上虽然是可选参数，但在实际开发中还是建议把每个模块的 controller 都指定好路径，避免混乱。

服务端如何读取请求对象？请求对象表示一个 HTTP 请求所携带的数据信息，如请求数据中的查询参数、路由参数、请求头、请求体等数据。表 6.1 列出的内置装饰器将简化请求数据信息的读取，并且满足了绝大部分业务的开发需求。

表 6.1　Nest 中常用的请求相关装饰器

装饰器	请求数据	装饰器	请求数据
@Request()、@Req()	req	@Body(key?: string)	req.body/req.body[key]
@Response()、@Res()	res	@Query(key?: string)	req.query/req.query[key]
@Next()	next	@Headers(name?: string)	req.headers/req.headers[name]
@Session()	req.session	@Ip()	req.ip
@Param(key?: string)	req.params/req.param[key]	@HostParam()	req.hosts

接下来我们在 OrdersController 控制器中编写更多的处理方法来演示接收不同的 HTTP 方法和不同位置的参数，通过装饰器的方式来读取。

① 如果有这样的一条 Get 请求：

```
curl --request GET \
--url 'http://localhost:3000/orders/list?page=1&limit=20'
```

在 Nest 中通过@Query 装饰器来读取，并在装饰器后声明参数就可以得到：

```
@Get('list')
list(@Query('page') page: number, @Query('limit') limit: number) {
  return `获取第${page}页，每页${limit}条订单`;
}
```

② 如果有这样的一条 Get 请求：

```
curl --request GET \
--url http://localhost:3000/orders/detail/1
```

通过 GET 方法查询指定 ID 的订单详情，并通过路由参数传递订单 ID。在 Nest 中通过@Param 就可以得到请求路由后的参数，就像这样：

```
@Get('detail/:id')
findById(@Param() param: { id: number }) {
  return `获取 ID 为 ${param.id} 的订单详情`;
}
```

③ 如果有这样的一条 Patch 请求：

```
curl --request PATCH \
--url 'http://localhost:3000/orders/1/FINSIHED'
```

通过 PATCH 方法更新指定 ID 订单的最新状态，并通过路由参数传递订单 ID 以及最新的状态。在 Nest 中，通过@Patch 来声明请求类型，通过@Param 来声明路由参数即可，就像这样：

```
@Patch(':id/:status')
updateByIdAndStatus(
  @Param('id') id: number,
  @Param('status') status: string,
) {
  return `将 ID 为 ${id} 订单状态更新为 ${status}`;
}
```

④ 如果有这样的一条 Post 请求：

```
curl --request POST \
--url http://localhost:3000/orders \
--header 'content-type: application/json' \
--data '{
"article": "HUAWEI-Meta60",
"price": 5999,
"count": 1,
"source": "Made in China"
}'
```

通过 POST 方法创建了一个新的订单，并且订单的数据通过请求体来传递。在 Nest 中，我们只需要接收一个参数，并且用@Body 装饰一下即可得到对应的请求体数据，就像这样：

```
interface ICreateOrder {
```

```
  article: string;
  price: number;
  count: number;
  source: string;
}

@Post()
create(@Body() order: ICreateOrder) {
  return `创建订单,订单信息为 ${JSON.stringify(order)}`;
}
```

看完这四个案例,我们简单来做个总结:
- 控制器中不同的处理函数可以通过 HTTP 方法来区分;
- 当多个处理函数需要使用相同的 HTTP 方法时,就需要添加处理函数级别的路由加以区分;
- 对于请求参数,@Query、@Param、@Body 可以得到对应指定的参数,满足大部分的业务场景。

了解了控制器(Controller)后,接下来我们开始学习提供者的使用,也就是 Providers。同样先看一张流程图,如图 6.4 所示。

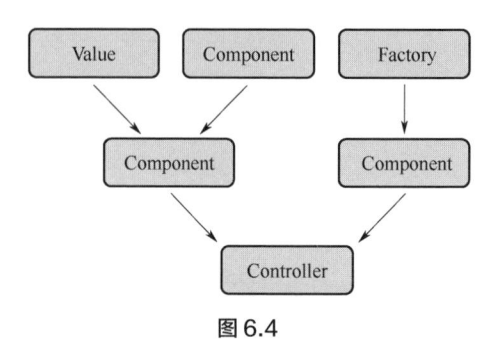

图 6.4

在 Nest.js 中,提供服务的类以及一些工厂类、助手类中所有提供能力的函数、类被称为提供者,它们均可以通过注入的方式作为依赖模块。快速创建服务的方式也很简单,通过 Nest.js 的 CLI 工具即可实现,同样有完整命令和简写命令两种:
- 完整命令: nest generate service orders
- 简写命令: nest g s orders

服务是很典型的提供者,HTTP 请求在经过控制器处理后进入的下一层就应该是服务层,也就是最为复杂的任务。整体的业务逻辑部分,如将复杂的订单生成、查询、更新以及删除等操作进行封装,就像这样:

```
import { Injectable } from '@nestjs/common';
```

```
import { CreateOrderDto } from './dto/create-order.dto';
import { UpdateOrderDto } from './dto/update-order.dto';

@Injectable()
export class OrdersService {
  create(createOrderDto: CreateOrderDto) {
    return 'This action adds a new order';
  }

  findAll() {
    return `This action returns all orders`;
  }

  findOne(id: number) {
    return `This action returns a #${id} order`;
  }

  update(id: number, updateOrderDto: UpdateOrderDto) {
    return `This action updates a #${id} order`;
  }

  remove(id: number) {
    return `This action removes a #${id} order`;
  }
}
```

上述代码是一个订单的基本服务,包含了增删改查,是最基本的服务组成部分。如果某个服务是通用的,可以选择注入其他服务中去使用,或者注入 OrderController 控制器中,这样就得到了初始化后的 OrderService 实例,就可以在不同的控制器处理函数中去调用该服务所提供的能力了,就像这样:

```
import { Controller, Get, Post, Body, Param } from '@nestjs/common';
import { OrdersService } from './orders.service';
import { CreateOrderDto } from './dto/create-order.dto';

@Controller('orders')
export class OrdersController {
  constructor(private readonly ordersService: OrdersService){}

  @Post()
  create(@Body() createOrderDto: CreateOrderDto) {
    return this.ordersService.create(createOrderDto);
```

```
  }

  @Get()
  findAll() {
    return this.ordersService.findAll();
  }

  @Get(':id')
  findOne(@Param('id') id: string) {
    return this.ordersService.findOne(+id);
  }
}
```

当开发接口时，将 controller 和 service 的关系保持上述代码这样即可，在 controller 中的接口被命中，最后一行代码永远都是调用对应的服务。当然除了构建函数注入这种方式以外，还可以使用属性注入，就像这样：

```
@Inject()
private readonly ordersService: OrdersService;
```

这也是比较推荐的方案，可以让代码看起来更直观。

学习完服务以后，我们再进入下一个概念的学习——中间件（middleware）。学习过前端的开发者其实都知道中间件的概念，如果使用过 axios，那肯定使用过请求拦截器或响应拦截器，它们与中间件概念一样，就是在某一个函数触发之前的拦截执行函数。中间件在 Nest.js 中的关系就像这样，如图 6.5 所示。

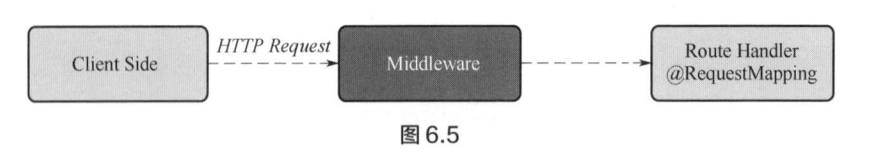

图 6.5

图 6.5 中很明显可以看到，网络请求在触达 middleware 之前是客户端的请求，middleware 通过之后就是 controller。由此，读者们应该大致清楚中间件的作用了，既然场景发生在接口触达前，那通常就是做权限校验的事情，如 JWT 验证用户身份信息这种事务。中间件是在路由处理程序前调用的函数，除了可以访问请求对象和响应对象以外，还提供 next()函数。

那我们如何创建中间件呢？与前面的介绍一致，用 nest 的 CLI 命令即可，同样提供了完整和简写两种命令，如下：

- 完整命令：nest g middleware logger
- 简写命令：nest g mi logger

代码中通常会在/src/middlewares 目录下，保存项目中所有的中间件，如果要开发一个打印日志的 logger 中间件，代码就像这样：

```
import { Injectable, NestMiddleware } from '@nestjs/common';

@Injectable()
export class LoggerMiddleware implements NestMiddleware {
  use(req: Request, res: Response, next: () => void) {
    console.log('Request...');
    next();
  }
}
```

当代码运行到 next()时，就会执行到下一个中间件，如果这个中间件是最后一个，就会运行到命中的 controller 中。那有了中间件如何生效呢？如何绑定消费者？中间件的使用方通常被称为消费，中间件和消费者（cats）的链接可以在 app 模块中进行处理，app 模块必须实现 NestModule 中的 configure()函数，并在这个函数中去完成关联。下面的代码是对于所有的接口都绑定 LoggerMiddleware 的例子：

```
export class AppModule implements NestModule {
  configure(consumer: MiddlewareConsumer) {
    consumer.apply(LoggerMiddleware).forRoutes('cats');
  }
}
```

上面案例是对所有路由产生消费，那如何进行显式路由匹配和排除呢？可以通过为 forRoutes 和 exclude 传入不同的参数实现中间件对路由范围的灵活控制，就像这样：

```
//基于模式匹配的应用方案
forRoutes({ path: 'ab*cd', method: RequestMethod.ALL });
```

或者是采用基于路由请求方法和路由名称的匹配方案，就像这样：

```
// 基于具体路由配置及模式匹配的排除方案
consumer
  .apply(LoggerMiddleware)
  .exclude(
    { path: 'cats', method: RequestMethod.GET },
    { path: 'cats', method: RequestMethod.POST },
    'cats/(.*)',
  )
  .forRoutes(CatsController);
```

而像我们写的 LoggerMiddleware 中间件，其功能简单，属于功能类中间件，没有额外的配置函数以及判断，也没有其他依赖关系，我们就可以直接声明全局中间件即可，就像这样：

```
const app = await NestFactory.create(AppModule);
app.use(logger);
await app.listen(3000);
```

上述代码中，中间件代表全局注册，它的消费者就是项目中所有的路由，将 app 模块中的接口以及接口实现移除，直接在 main.ts 中 app 实例化后进行 use 函数注册即可。

学习完中间件，我们再学习下一个比较重要的概念——异常过滤器。当我们的 JavaScript 代码出现异常时，会导致项目报错，对于 Nest.js 服务层也是如此，并且报错以后整个服务会失效，因此在适当的地方进行 try catch 拦截并抛出对应的异常就是过滤器的作用。先看一张官方对于异常过滤器的图解，如图 6.6 所示。

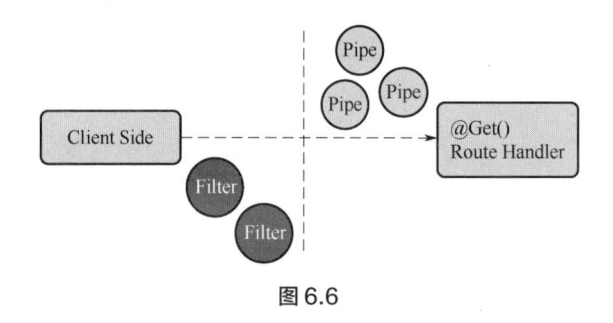

图 6.6

异常层由开箱即用的全局异常过滤器来处理，负责处理整个应用程序中所有未处理的异常。通过内置的 HttpException 类可以轻松抛出一个标准的异常，就像这样：

```
@Get('find')
findCatById(@Query('id') id: string): Cat | undefined {
  try {
    // TODO
  } catch (error) {
    throw new HttpException('Forbidden', HttpStatus.FORBIDDEN, {
      cause: error,
    });
  }
  return this.catsService.findCatById(Number(id));
}
```

在触发异常后客户端将收到一份 JSON 格式的报文。cause 作为可选项虽然不会序列化后发送到客户端，但可以作为日志记录使用。返回给客户端的异常报文如下：

```
{
  "statusCode": 403,
  "message": "Forbidden"
}
```

当然也可以自定义异常，直接使用内置的 HttpException 实现标准异常的抛出，再基于 HttpException 继承一层就可以，就像下面这段代码实现了一个 ForbiddenException 异常类：

```
import { HttpException, HttpStatus } from '@nestjs/common';

export class ForbiddenException extends HttpException {
  constructor(error: unknown) {
    super('Forbidden', HttpStatus.FORBIDDEN, {
      cause: error,
    });
  }
}
```

Nest 中内置了非常多的 HTTP 异常，都继承于 HttpException 类，它们可以满足绝大部分的业务开发场景，非必要时不需要自己再去封装。

那我们应该如何创建一个异常过滤器呢？和之前的方法一样，利用 Nest.js 提供的 CLI 命令创建即可，就像这样：

- 完整命令：nest g filter http-exceptionhuo
- 简写命令：nest g f http-exception

这样就可以在 Nest 应用中快速创建一个用来接管内置异常过滤器的指定过滤器了，通过重写 catch() 可以实现具体的拦截处理。catch() 方法的参数中，exception 参数是当前正在处理的异常对象；host 参数是一个 ArgumentsHost 对象，从 host 参数获取对传递给原始请求处理程序（在异常产生的控制器中）的 Request 和 Response 对象的引用。下面是一个封装异常过滤器的示例代码：

```
import {
  ArgumentsHost,
  Catch,
  ExceptionFilter,
  HttpException,
} from '@nestjs/common';
```

```
@Catch(HttpException)
export class HttpExceptionFilter implements ExceptionFilter {
  catch(exception: HttpException, host: ArgumentsHost) {
    const ctx = host.switchToHttp();
    const resp = ctx.getResponse();
    const req = ctx.getRequest();
    const status = exception.getStatus();
    resp.status(status).json({
      statusCode: status,
      timestamp: new Date().toISOString(),
      path: req.url,
    });
  }
}
```

当异常发生时，我们返回了错误的状态码 statusCode、发生异常的时间戳 timestamp 以及请求接口路由 path。那么我们该如何使用封装好的异常拦截器呢？可以通过@UseFilters(HttpExceptionFilter)绑定到需要拦截的控制器处理函数，就像这样：

```
@Get('find')
@UseFilters(HttpExceptionFilter)
findCatById(@Query('id') id: string): Cat | undefined {
  return this.catsService.findCatById(Number(id));
}
```

或者也可以直接绑定到整个控制器的上层，这样所有的接口都会生效，就像这样：

```
@UseFilters(HttpExceptionFilter)
@Controller('cats')
export class CatsController {
  //TODO
}
```

如果需要全局使用这个异常拦截器，那么就直接在 app 实例化之后通过 useGlobalFilters()函数进行设置，就像这样：

```
async function bootstrap() {
  const app = await NestFactory.create(AppModule);
  app.useGlobalFilters(new HttpExceptionFilter());
  await app.listen(3000);
}
```

```
bootstrap();
```

当然如果需要在模块级别设置异常过滤器，则可以这么做：

```
@Module({
controllers: [CatsController],
providers: [
  CatsService,
  // 设置异常过滤器
  {
    provide: APP_FILTER,
    useClass: HttpExceptionFilter,
  },
],
  exports: [CatsService],
})
export class CatsModule {}
```

前面的异常过滤器我们都只是用于 HttpException 的异常拦截，如果过滤器想要拦截包含除 HttpException 以外的所有异常，就需要进行进一步的处理和适配。我们快速创建一个全局异常过滤器（nest g f all-exceptions），并注入 HttpAdapterHost 适配器来处理异常情况：

```
import {
ArgumentsHost,
Catch,
ExceptionFilter,
HttpException,
HttpStatus,
} from '@nestjs/common';
import { AbstractHttpAdapter } from '@nestjs/core';

@Catch()
export class AllExceptionsFilter implements ExceptionFilter {
  constructor(private readonly httpAdapter: AbstractHttpAdapter) {}
  catch(exception: unknown, host: ArgumentsHost): void {
    const ctx = host.switchToHttp();
    const httpStatus =
      exception instanceof HttpException
      ? exception.getStatus()
      : HttpStatus.INTERNAL_SERVER_ERROR;
    const responseBody = {
      statusCode: httpStatus,
```

```
      timestamp: new Date().toISOString(),
      path: this.httpAdapter.getRequestUrl(ctx.getRequest()),
    };
    this.httpAdapter.reply(ctx.getResponse(), responseBody, httpStatus);
  }
}
```

上述代码中，我们对于 httpStatus、path 进行了兼容判断，并通过 httpAdapter 对象来完成适配，接下来将它绑定到 app 实例上：

```
import { HttpAdapterHost, NestFactory } from '@nestjs/core';
import { AppModule } from './app.module';
import { AllExceptionsFilter } from './all-exceptions/all-exceptions.filter';

async function bootstrap() {
  const app = await NestFactory.create(AppModule);
  const { httpAdapter } = app.get(HttpAdapterHost);
  app.useGlobalFilters(new AllExceptionsFilter(httpAdapter));
  await app.listen(3000);
}
bootstrap();
```

从 app 模块中得到 httpAdapter 对象并传入异常拦截器，从而实现进一步的异常类型兼容。

接下来，再来了解一个新的概念——管道。先看一张管道的后端交互图，如图 6.7 所示。

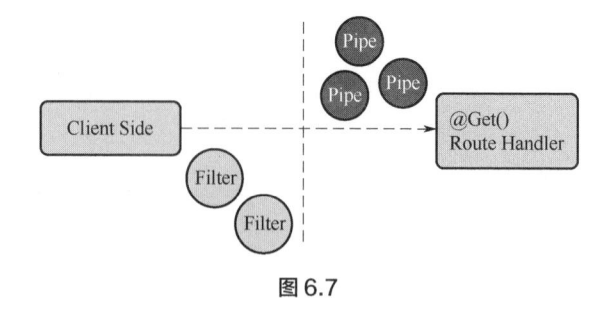

图 6.7

管道在 Nest.js 中提供转换（将输入数据转换为所需的形式）、验证（输入数据是否有效，有效则向下传递，否则抛出异常）两大类功能。Nest.js 提供了一些内置管道，如 ValidationPipe、DefaultValuePipe、ParseIntPipe 等。

接下来我们开始尝试绑定管道：

```
@Get('find')
findCatById(@Query('id') id: number): Cat | undefined {
  return this.catsService.findCatById(id);
}
```

上面的代码中，控制器处理函数虽然声明为 number 类型，但 typeof ID 仍然收到了一个 string 类型的数据，这样的数据传递到服务层处理是很危险的，现在我们尝试绑定 ParseIntPipe 来解决这个问题。绑定 ParseInitPipe 管道到 findCatById 处理函数，当路由到此处理函数时，ParseInitPipe 管道将尝试解析 ID 数据 number 类型，解析成功就会正常调用服务层逻辑，解析失败将触发异常(Validation failed (numeric string is expected))，代码如下：

```
@Get('find')
findCatById(@Query('id', ParseIntPipe) id: number): Cat | undefined {
  return this.catsService.findCatById(id);
}
```

在绑定管道的时候还可以传递管道实例，通过管道的构造函数提供的选项进行定制化，就像这样：

```
@Get('find')
findCatById(
  @Query(
    'id',
    new ParseIntPipe({
      errorHttpStatusCode: HttpStatus.NOT_ACCEPTABLE,
    }),
  )
  id: number,
): Cat | undefined {
  return this.catsService.findCatById(id);
}
```

介绍完系统内置的管道，接下来按照惯例继续看一下自定义管道，同样用 Nest 的 CLI 命令即可，命令如下：

- 完整命令：nest g pipe validation
- 简写命令：nest g pi validation

我们创建一个 ValidationPipe 管道，并且加入两个日志，代码如下：

```
@Injectable()
export class ValidationPipe implements PipeTransform {
  transform(value: any, metadata: ArgumentMetadata) {
```

```
    console.log('value', value); // 2
    console.log('metadata', metadata); // { metatype: [Function: Number],
    type: 'query', data: 'id' }
    return value;
  }
}
```

Nest 对于管道中的 transform 方法返回了两个参数：

- value：处理函数的参数，当请求发送的 ID 为 1 时，value 将打印为 1。
- metadata：处理函数参数的元数据，type 表示参数来自 Body、Query、Param，还是自定义参数；data 表示传递给装饰器的值；metatype 表示提供参数的元类型。

介绍完基本数据类型以后，对象应该如何验证呢？这里我们需要引用 Zod 模块，使用其提供的可读的 API，以简单的方式来创建模式，并完善管道，执行安装命令：

```
npm install --save zod
```

然后在自定义拦截器中引入 Zod 模块，并通过 parse 方法来解析，代码如下：

```
import { BadRequestException, Injectable, PipeTransform } from '@nestjs/common';
import { ZodObject } from 'zod';

@Injectable()
export class ValidationPipe implements PipeTransform {
  constructor(private schema: ZodObject<any>) {}

  transform(value: unknown) {
    try {
      this.schema.parse(value);
    } catch (error) {
      throw new BadRequestException('Validation failed');
    }
    return value;
  }
}
```

然后，基于入参的对象类型创建一个 Zod 模块的验证对象，代码如下：

```
import { z } from 'zod';

export const createCatSchema = z
  .object({
```

```
    name: z.string(),
    age: z.number(),
  })
  .required();

export type CreateCatDto = z.infer<typeof createCatSchema>;
```

最后,将更新后的验证管道使用@UsePipes 装饰器绑定到业务处理函数上,代码如下:

```
@Post('create')
@UsePipes(new ValidationPipe(createCatSchema))
create(@Body() cat: Cat): Cat[] | undefined {
  return this.catsService.create(cat);
}
```

那么,应该如何对 Class 类进行验证呢?我们可以选择使用装饰器对 Class 的属性进行表述来实现基于 Class 的验证。同样还是先执行命令安装依赖包:

```
npm i --save class-validator class-transformer
```

接着,从 class-validator 模块导入 IsString 和 isInt 装饰器,来验证类中的基本数据类型:

```
import { IsString, IsInt } from 'class-validator';

export class Cat {
  @IsInt()
  id: number;

  @IsString()
  name: string;

  @IsInt()
  age: number;
}
```

回顾一下管道的学习,我们掌握了基本数据类型 String、Number 的验证,又掌握了对象和类的验证,现在我们将管道重构,让它可以按顺序验证多个数据类型:

```
import {
  ArgumentMetadata,
  BadRequestException,
  Injectable,
```

```
  PipeTransform,
} from '@nestjs/common';
import { plainToInstance } from 'class-transformer';
import { validate } from 'class-validator';

@Injectable()
export class ValidationPipe implements PipeTransform<any> {
  async transform(value: any, { metatype }: ArgumentMetadata) {
    // ① 初筛，处理预期内的数据类型
    if (!metatype || !this.toValidate(metatype)) {
      return value;
    }

    // ② 将 value 和元类型转为实例对象
    const object = plainToInstance(metatype, value);
    // ③ 通过 validate 验证结果
    const errors = await validate(object);
    if (errors.length > 0) {
      throw new BadRequestException('Validation failed');
    }
    return value;
  }

  private toValidate(metatype: any): boolean {
    const types: any[] = [String, Boolean, Number, Array, Object];
    return !types.includes(metatype);
  }
}
```

这样自定义管道就重构好了，基于 String、Boolean、Number、Array、Object 的类型，将直接进行初筛。对于类，再基于 class-transformer、class-validator 来进行验证。

最后，我们要将这个完善的自定义管道作用绑定到全局，和之前的逻辑一样，在 app 实例化后调用一次即可，就像这样：

```
async function bootstrap() {
  const app = await NestFactory.create(AppModule);
  app.useGlobalPipes(new ValidationPipe());
  await app.listen(3000);
}
bootstrap();
```

如果采用依赖注入方案,借助任何模块外注册的全局管道[例如上述示例中的 useGlobalPipes()]将无法正常发挥作用,因为绑定过程是在所有模块上下文之外进行的。为了解决这一问题,可以使用以下构造方法来配置全局管道,从而确保有效的依赖注入和注册:

```
import { Module } from '@nestjs/common';
import { APP_PIPE } from '@nestjs/core';

@Module({
  providers: [
    {
      provide: APP_PIPE,
      useClass: ValidationPipe,
    },
  ],
})
export class AppModule {}
```

至此,在 Nest.js 中使用管道的相关知识都已经讲解完了。到这里,还剩下最后一个概念——Nest.js 中的拦截器。熟悉前端的开发者对于这个词一定不陌生,例如我们封装请求库时喜欢使用 Axios 库,可以在 Axios 中配置请求拦截器、响应拦截器,从而实现在请求之前添加额外的参数,例如 Header、token;也可以在收到响应后捕获异常进行埋点的上报、客户端异常的透出以及其他兜底处理。

在 Nest.js 中,拦截器是一个 APO 切面编程技术,应用拦截器可以获得下面所列出的一系列能力,其实和前端是类似的:

- 在方法执行之前/之后绑定额外的逻辑;
- 转换函数返回的结果;
- 转换函数抛出的异常;
- 扩展基本功能行为;
- 根据特定条件完全覆盖函数,例如接口缓存的能力。

例如,可以通过拦截器来统计处理函数执行时间,即使用拦截器在不侵入处理函数的前提下计算处理函数执行的时长,这是一个典型的切面编程案例。和之前的案例一样,现在使用 CLI 命令 nest g interceptor logging 或简写命令 nest g itc logging 来创建 logging 拦截器:

```
import {
  CallHandler,
  ExecutionContext,
  Injectable,
```

```
  NestInterceptor,
} from '@nestjs/common';
import { Observable, tap } from 'rxjs';

@Injectable()
export class LoggingInterceptor implements NestInterceptor {
  intercept(context: ExecutionContext, next: CallHandler):
  Observable<any> {
    console.log('Before...');
    const now = Date.now();
    return next
      .handle()
      .pipe(tap(() => console.log(`After... ${Date.now() - now}ms`)));
  }
}
```

在拦截器中我们使用了 rxjs 技术，tap 运算符将在处理函数执行结束后计算所执行的时间。绑定拦截器也是通过装饰器注入的形式，可以在局部控制器中注入，就像这样：

```
@UseInterceptors(LoggingInterceptor)
export class CatsController {}

//or

@UseInterceptors(new LoggingInterceptor())
export class CatsController {}
```

或者是在 app 实例化之后，在全局范围中绑定，就像这样：

```
const app = await NestFactory.create(AppModule);
app.useGlobalInterceptors(new LoggingInterceptor());

// or

@Module({
  providers: [
    {
      provide: APP_INTERCEPTOR,
      useClass: LoggingInterceptor,
    },
  ],
})
```

```
export class AppModule {}
```

我们还可以通过 rxjs 提供的 map 操作符对处理函数返回的数据进行二次加工，比如可以把执行的时间一起返回，便于统计接口性能，就像这样：

```
@Injectable()
export class TransformInterceptor implements NestInterceptor {
  intercept(context: ExecutionContext, next: CallHandler):
  Observable<any> {
    return next.handle().pipe(
      map((data) => {
        return {
          time: new Date().toISOString(),
          data,
        };
      }),
    );
  }
}
```

我们也可以基于 rxjs 提供的 catchError 操作符来抛出指定的异常：

```
@Injectable()
export class ErrorsInterceptor implements NestInterceptor {
  intercept(context: ExecutionContext, next: CallHandler):
  Observable<any> {
    return next
      .handle()
      .pipe(catchError((err) => throwError(() => new
    BadGatewayException())));
  }
}
```

最后还有一个比较常见的场景，我们可以通过 rxjs 提供的 timeout 和 catchError 来实现接口执行超时异常抛出的能力，就像这样：

```
@Injectable()
export class TimeoutInterceptor implements NestInterceptor {
  intercept(context: ExecutionContext, next: CallHandler):
  Observable<any> {
    return next.handle().pipe(
      timeout(5 * 1000),
      catchError((err) => {
        if (err instanceof TimeoutError) {
```

```
            return throwError(() => new RequestTimeoutException());
          }
          return throwError(() => err);
        }),
      );
    }
  }
```

至此，对于 Nest.js 或者说是后端的常见概念，相信读者也已经有所了解了。在本小节的最后，做如下小结：
- 使用@nestjs/cli 创建项目及模块；
- 控制器的使用：处理每次客户端的请求；
- 服务的使用：封装复杂的业务逻辑，并提供此能力给其他模块；
- 模块的使用：负责项目所有控制器、提供者的连接管理工作；
- 中间件的使用：更改请求的响应对象以及执行下一个中间件；
- 异常过滤器的使用：处理项目中所有未处理的异常；
- 管道的使用：对于客户端传入的数据进行转换和验证；
- 拦截器的使用：对处理函数进行切面上的扩展。

6.2 数据库连接和初始化

在上一节介绍 Nest.js 中相关概念的基础上，从本节开始，将以案例形式引导读者了解其应用。Nest.js 作为后端框架，需要数据库的支持，那么集成数据库前需要做哪些准备工作？

首先，我们需要安装和构建数据库表相关的 npm 包：
- Typescript：作为微软开发的编程语言，它是 JavaScript 的一个超集，添加了静态类型支持。使用 TypeScript 可以在开发过程中提供更强的类型检查，帮助开发者在编码阶段发现潜在的错误。在 Nest.js 项目中，TypeScript 通常用于编写应用程序的代码。
- typeorm：一个支持多种数据库的对象关系映射（ORM）库，它允许使用 TypeScript 或 JavaScript 语言编写数据库相关的代码，而不需要直接操作 SQL 语句。typeorm 可以帮助开发者在数据库表和应用程序的数据模型之间建立映射关系，便捷地进行数据库增、删、改、查操作。
- class-validator：基于类装饰器和方法装饰器的验证库，它可以用于在 Nest.js 应用中验证输入数据的合法性。通过 class-validator，开发者可以方便地在 DTO（数据传输对象）或实体类上定义验证规则，确保数据的有效性和安全性。

- @nestjs/typeorm：Nest.js 官方提供的一个 typeorm 模块，用于集成 typeorm 到 Nest.js 应用中，简化了在 Nest.js 中使用 typeorm 的配置和使用，提供了一些装饰器和依赖注入的能力，使得在 Nest.js 中进行数据库操作更加便捷。
- mysql2：作为 mysql 数据库的 Node.js 驱动程序，它提供了异步、非阻塞的 mysql 数据库连接和查询功能。在 Node.js 应用中，mysql2 可以用于连接 mysql 数据库服务器，并执行查询、更新、删除等数据库操作。

这五大套件通常在 Nest.js 项目中一起使用，笔者所用依赖包如图 6.8 所示，用于构建具有数据库支持的 Web 应用程序。Typescript 用于编写类型安全的代码；typeorm 和 @nestjs/typeorm 用于数据库操作；class-validator 用于验证用户输入；mysql2 作为 mysql 数据库的 Node.js 驱动程序。

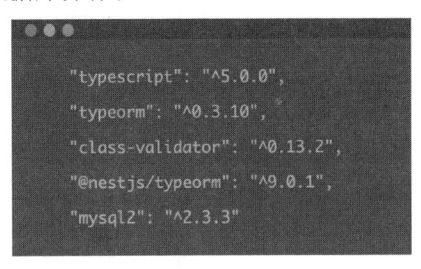

图 6.8

对于项目结构，Nest.js 采用模块化架构，而所有的数据库表都会存放在 /src/entitys 目录中，接下来我们创建一个 user.entity.ts，也就是 user 用户表。整个 user 表代码如下：

```
import { IsNotEmpty } from 'class-validator';
import { Column, Entity, PrimaryGeneratedColumn } from 'typeorm';

/**
 * 组件
 */
@Entity('user')
export class User {
  @PrimaryGeneratedColumn('uuid')
  id: string;

  @Column({
    comment: '昵称',
    default: '',
  })
  @IsNotEmpty()
  name: string;

  @Column({
    comment: '描述',
    default: '',
  })
```

```
  desc: string;

  @Column({
    comment: '手机号',
    nullable: true,
  })
  tel: string;

  @Column({
    comment: '密码',
    nullable: true,
  })
  password: string;

  @Column({
    comment: '账户',
    nullable: true,
  })
  account: string;
}
```

在代码中，我们使用了 Nest.js 中的装饰器和 typeorm 中的实体（Entity）装饰器，用于定义一个名为 User 的实体类，在这个类中，使用了装饰器来添加元数据和验证规则。

首先我们导入了 class-validator 中的 IsNotEmpty 装饰器，用于定义属性的验证规则。同时，还导入了 typeorm 中的 Column、Entity 和 PrimaryGeneratedColumn 装饰器，用于定义数据库表的结构和字段。

其中@Entity("user")是 typeorm 中的实体装饰器，用于指定实体在数据库中的表名为 user，需要和实际库中的命名保持一致。

export class User{}定义了名为 User 的类，表示数据库中的用户实体，可以在 Service 中以注入的形式消费。

在 typeorm 中，@Entity 是一个装饰器，用于将一个类标记为一个数据库实体（Entity）。数据库实体通常对应数据库中的一张表，每个实体类的实例代表了表中的一行数据。使用@Entity 装饰器，开发者可以告诉 typeorm 将这个类映射到数据库中的某个表。

在给@Entity 装饰器传递参数时，开发者可以指定实体在数据库中的表名。@Entity("user")中的 user 参数指定了这个实体在数据库中的表名为 user。typeorm 将会使用这个表名来创建数据库表，并且将 User 类的实例映射到这个表中的数据记录中。

如果不指定表名，在默认情况下，typeorm 将使用类名作为数据库表的名称。其次是定义实体中的字段属性。@PrimaryGeneratedColumn("uuid")id: string 通过该装饰器定义了一个自动生成的护肩 ID，类型为 string，使用 uuid 数据类型。

@Column 昵称字段使用该装饰器生成了一个普通字段，指定该数据库中的注释为昵称，默认值为空字符，并且使用了 @IsNotEmpty() 装饰器表示该字段不可为空。

在 Nest.js 中完成数据表实体的创建后，接下来如何在模块中使用该实体呢？在 module 文件中，基于 @nestjs/typeorm 库提供的 TypeormModule API 即可实现，就像这样：

```
import { Module, ConsoleLogger } from '@nestjs/common';
import { TypeOrmModule } from '@nestjs/typeorm';
import { User } from './models/user.entity';

@Module({
  imports: [TypeOrmModule.forFeature([User])],
  providers: [ConsoleLogger],
  exports: [],
})
export class UserModule {}
```

至此，我们已经完成了数据表的创建和导入使用，接下来我们需要将准备好的数据库与 Nest 应用建立连接，在 App 模块中全局连接导入即可，还是使用 TypeormModule 方法，代码如下：

```
import { Module } from '@nestjs/common';
import { AppController } from './app.controller';
import { AppService } from './app.service';
import { TypeOrmModule } from '@nestjs/typeorm';
import { UserModule } from './modules/user/user.module';

@Module({
  imports: [
    TypeOrmModule.forRoot({
      type: 'mysql',
      host: 'X.X.X.X',
      port: 3306,
      username: 'root',
      password: '123456',
      database: 'healthflex',
```

```
      entities: [`${__dirname}/../modules/**/*.entity{.ts,.js}`],
      logging: true,
      synchronize: true,
      autoLoadEntities: true,
    }),
    UserModule,
  ],
  controllers: [AppController],
  providers: [AppService],
})
export class AppModule {}
```

TypeOrmModule.forRoot()方法用于配置数据库连接。该方法接收一个配置对象作为参数，该配置对象包含了连接数据库所需的各种信息：

- type：指定数据库的类型，例如 mysql、postgres、splite。
- host：数据库服务器的主机地址。
- port：数据库服务器的主机地址。
- username：连接数据库的用户名。
- password：连接数据库的密码。
- database：要连接的数据库名称。
- entities：指定实体类，即所有的 entity.ts 文件，例如上例的 User 实体。
- synchronize：设置为 true 时，每次应用启动时都会根据实体类自动创建数据库表。在开发环境中，可以使用这个选项方便地创建数据库表；在生产环境中，通常设置为 false，避免应用程序自动修改数据库结构。
- logging：是否启用数据库日志记录。在开发环境中，可以设置为 true 以便于调试和查看数据库操作记录；在生产环境中，通常设置为 false，以提高数据库性能。

随后运行 Nest 项目，基于 synchronize 自动建表，我们最早创建的 User 表自动创建好了，如图 6.9 所示。

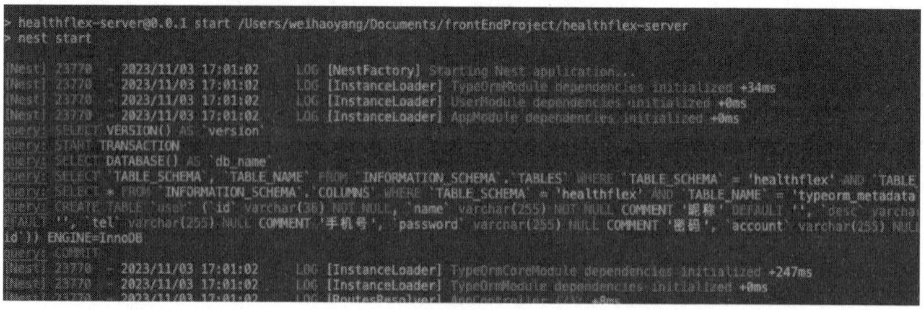

图 6.9

6.3 快速构建 CRUD

一个全新的后端框架避免不了 CRUD 的实现，就像前端避免不了 TODO-LIST 一样。本节我们来实现一下基于 Nest.js+typeorm+mysql2 的用户 CRUD 案例。

首先创建 Nest 项目，执行命令：

```
nest new nest-user-crud
```

创建完项目后，我们安装数据库套件，执行依赖安装命令：

```
npm install --save @nestjs/typeorm typeorm mysql2
```

然后，我们需要配置 typeorm 以连接到 mysql 数据库，在项目根目录创建一个 ormConfig.json 文件，添加以下配置：

```
{
  "type": "mysql",
  "host": "localhost",
  "port": 3306,
  "username": "root",
  "password": "your_password",
  "database": "nest_user_crud",
  "entities": ["dist/**/*.entity{.ts,.js}"],
  "synchronize": true
}
```

将这部分代码导入 App 模块中让项目与数据库关联起来。

```
@Module({
  imports: [
    TypeOrmModule.forRoot(ormConfig),
    UserModule,
  ],
  controllers: [AppController],
  providers: [AppService],
})
export class AppModule {}
```

接下来我们创建一个新的模块来进行 CRUD 设计。执行命令创建一个 user 模块：

```
nest g module users
```

然后我们在模块中创建一个新的 user.entity.ts 文件，定义 User 实体，代码如下：

```
import { Entity, PrimaryGeneratedColumn, Column } from 'typeorm';
```

```
@Entity()
export class User {
  @PrimaryGeneratedColumn()
  id: number;

  @Column()
  username: string;

  @Column()
  password: string;
}
```

接下来,需要创建对应的 user 服务来处理所有用户实体相关的业务逻辑,运行以下命令:

```
nest g service users
```

然后在 user.service.ts 文件中,实现基本的 CRUD 操作,代码如下:

```
import { Injectable } from '@nestjs/common';
import { InjectRepository } from '@nestjs/typeorm';
import { Repository } from 'typeorm';
import { User } from './user.entity';

@Injectable()
export class UsersService {
  constructor(
    @InjectRepository(User)
    private usersRepository: Repository<User>,
  ) {}

  async create(user: User): Promise<User> {
    return this.usersRepository.save(user);
  }

  async findAll(): Promise<User[]> {
    return this.usersRepository.find();
  }

  async findOne(id: number): Promise<User> {
    return this.usersRepository.findOne(id);
  }
```

```typescript
  async update(id: number, user: User): Promise<void> {
    await this.usersRepository.update(id, user);
  }

  async remove(id: number): Promise<void> {
    await this.usersRepository.delete(id);
  }
}
```

最后，把对应 user 的控制器 controller 创建出来，来实现基于 service 的 CRUD 客户端接口，运行以下命令：

```
nest g controller users
```

在 user.controller.ts 文件中编码，应遵循一对一的原则，即 1 个 controller 函数对应 1 个 service，就像这样：

```typescript
import { Controller, Get, Post, Put, Delete, Body, Param } from '@nestjs/common';
import { UsersService } from './users.service';
import { User } from './user.entity';

@Controller('users')
export class UsersController {
  constructor(private readonly usersService: UsersService) {}

  @Post()
  async create(@Body() user: User): Promise<User> {
    return this.usersService.create(user);
  }

  @Get()
  async findAll(): Promise<User[]> {
    return this.usersService.findAll();
  }

  @Get(':id')
  async findOne(@Param('id') id: number): Promise<User> {
    return this.usersService.findOne(id);
  }

  @Put(':id')
  async update(@Param('id') id: number, @Body() user: User): Promise<void> {
```

```
    return this.usersService.update(id, user);
  }

  @Delete(':id')
  async remove(@Param('id') id: number): Promise<void> {
    return this.usersService.remove(id);
  }
}
```

至此，我们便基于 Nest.js 实现了用户 CRUD 操作的所有后端逻辑。从实际案例中，或许读者也能感受到基于 JavaScript 开发后端的便捷性。

6.4 基于 Nest.js 的 RBAC 权限控制系统实现

什么是 RBAC 系统？RBAC（role-based access control，基于角色的访问控制）是一种权限控制机制，是在系统中按照用户的不同角色来分配权限的方法。在 RBAC 模型中，访问权限不是直接分配给每个用户，而是与用户的角色相关联，用户通过其角色获得相应的权限。

RBAC 模型主要包括了以下几个核心概念：

- 用户（User）：系统中的实际操作者，如员工或客户；
- 角色（Role）：一组权限的集合，通常与工作职责相关，例如"管理员""编辑""游客"等；
- 权限（Permission）：权限是对于系统中资源的访问和操作所需的授权，例如读取、编辑、删除文件等操作，是最小颗粒度概念；
- 会话（Session）：用户与系统交互的时间期限，在这个时间内用户有权访问系统。

系统管理员会将用户分配给一个或多个角色，每个角色拥有特定的权限集合。例如，普通用户能看到 A、B 页面；高级用户可以看到 A、B、C、D 页面；超级管理员可以看到所有的页面。当用户登录系统时，他们会获得与其角色相关联的权限。

RBAC 可以简化权限管理，因为管理员可以通过设置角色来管理一组用户的权限，而不需要单独为每个用户分配权限。这种业务的逻辑背后就像是如图 6.10 所示的关系。

接下来就基于 Nest.js 来从零到一实现

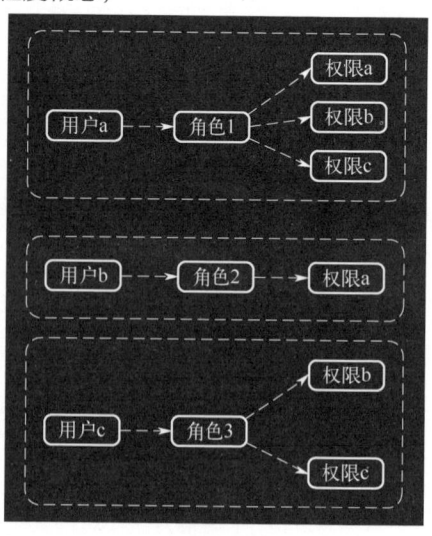

图 6.10

系统根基——权限设计。

① 新建一个 Nest 项目，执行以下命令：

```
nest new nest-project
```

② 安装数据库的相关依赖，主要是 typeorm、mysql2，在项目根目录执行以下命令：

```
npm install --save @nestjs/typeorm typeorm mysql2
```

③ 进行 typeorm 的相关配置，配置在 app.module.ts 文件中，代码如下：

```
import { Module } from '@nestjs/common'
import { AppController } from './app.controller'
import { AppService } from './app.service'
import { TypeOrmModule } from '@nestjs/typeorm'

@Module({
  imports: [
    TypeOrmModule.forRoot({
      type: 'mysql',
      host: 'localhost',
      port: 3306,
      username: 'root',
      password: 'password',
      database: 'nest-database',
      synchronize: true,
      logging: true,
      entities: [__dirname + '/**/*.entity{.ts,.js}'],
      poolSize: 10,
      connectorPackage: 'mysql2'
    })
  ],
  controllers: [AppController],
  providers: [AppService]
})

export class AppModule {}
```

④ 进行表设计，一般 RBAC 系统会有 5 张表，如下：
- 用户表：user，保存用户基本信息，如用户名、密码、邮箱；
- 角色表：role，保存角色基本信息，如角色名、角色 code；
- 权限表：permission，保存权限基本信息，如权限名、权限 code；

- 用户角色关联表：user_role_relation，用于记录用户和角色的关系；
- 角色权限关联表：role_permission_relation，用于记录角色和权限的关系。

领域模型设计图如图 6.11 所示。

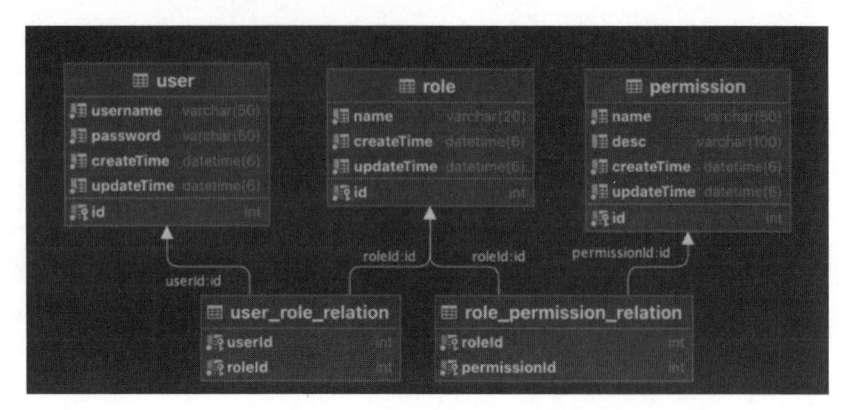

图 6.11

⑤ 在项目中新建三张非关联表，并把关联的关系记录下来。

首先是用户表 user.entity.ts，代码如下：

```
import {
  Column,
  CreateDateColumn,
  Entity,
  JoinTable,
  ManyToMany,
  PrimaryGeneratedColumn,
  UpdateDateColumn
} from 'typeorm'

import { Role } from './role.entity'

@Entity()
export class User {
  @PrimaryGeneratedColumn()
  id: number

  @Column({
    length: 50
  })
  username: string
```

```
  @Column({
    length: 50
  })
  password: string

  @CreateDateColumn()
  createTime: Date

  @UpdateDateColumn()
  updateTime: Date

  @ManyToMany(() => Role)
  @JoinTable({
    name: 'user_role_relation',
    joinColumn: {
      name: 'userId',
      referencedColumnName: 'id',
    },
    inverseJoinColumn: {
      name: 'roleId',
      referencedColumnName: 'id',
    },
  })
  roles: Role[];
}
```

这一步需重点关注在 user 表中定义的 rule 字段。通过连接到 user_role_relation 表，我们实现了关联关系的建立。只有在 user.id === userRoleRelation.userId 且 role.id === userRoleRelation.roleId 的情况下，才会将符合条件的 Role 记录插入到 User 中，从而自动形成关联。

然后，再初始化 role 角色表的实体文件 role.entity.ts，代码如下：

```
import {
  Column,
  CreateDateColumn,
  Entity,
  JoinTable,
  ManyToMany,
  PrimaryGeneratedColumn,
  UpdateDateColumn
} from 'typeorm'
import { Permission } from './permission.entity'
```

```ts
@Entity()
export class Role {
  @PrimaryGeneratedColumn()
  id: number

  @Column({
    length: 20
  })
  name: string

  @CreateDateColumn()
  createTime: Date

  @UpdateDateColumn()
  updateTime: Date

  @ManyToMany(() => Permission)
  @JoinTable({
    name: 'role_permission_relation',
    joinColumn: {
      name: 'roleId',
      referencedColumnName: 'id',
    },
    inverseJoinColumn: {
      name: 'permissionId',
      referencedColumnName: 'id',
    },
  })
  permissions: Permission[]
}
```

这张角色表中的 permission 字段与前述原理相同。通过连接到 role_permission_relation 表，实现了关联关系的建立。只有当 role.id === rolePermissionRelation.roleId 且 permission.id === rolePermissionRelation.permissionId 时，才会将符合条件的 Permission 记录插入到 Role 表中，从而自动形成关联。

最后是权限表 permission.entity.ts。其没有关联关系，直接记录有哪些权限即可，代码如下：

```ts
import {
  Column,
  CreateDateColumn,
```

```
  Entity,
  PrimaryGeneratedColumn,
  UpdateDateColumn,
} from 'typeorm';

@Entity()
export class Permission {
  @PrimaryGeneratedColumn()
  id: number;

  @Column({
    length: 50,
  })
  name: string;

  @Column({
    length: 100,
    nullable: true,
  })
  desc: string;

  @CreateDateColumn()
  createTime: Date;

  @UpdateDateColumn()
  updateTime: Date;
}
```

到这里，表的设计就完成了，我们写一个接口来初始化一批数据，即写一个 init 服务，生成两个用户，给这些用户套上一些角色，并且创建一批角色和权限，做一些关联操作，代码如下：

```
async initData() {
  const user1 = new User()
  user1.username = '张三'
  user1.password = '111111'

  const user2 = new User()
  user2.username = '李四'
  user2.password = '222222'

  const user3 = new User()
  user3.username = '王五'
```

```
user3.password = '333333'

const role1 = new Role()
role1.name = '管理员'

const role2 = new Role()
role2.name = '普通用户'

const permission1 = new Permission()
permission1.name = '新增 aaa'

const permission2 = new Permission()
permission2.name = '修改 aaa'

const permission3 = new Permission()
permission3.name = '删除 aaa'

const permission4 = new Permission()
permission4.name = '查询 aaa'

const permission5 = new Permission()
permission5.name = '新增 bbb'

const permission6 = new Permission()
permission6.name = '修改 bbb'

const permission7 = new Permission()
permission7.name = '删除 bbb'

const permission8 = new Permission()
permission8.name = '查询 bbb'

role1.permissions = [
  permission1,
  permission2,
  permission3,
  permission4,
  permission5,
  permission6,
  permission7,
  permission8
]
```

```
  role2.permissions = [permission1, permission2, permission3, permission4]

  user1.roles = [role1]

  user2.roles = [role2]

  await this.entityManager.save(Permission, [
    permission1,
    permission2,
    permission3,
    permission4,
    permission5,
    permission6,
    permission7,
    permission8
  ])

  await this.entityManager.save(Role, [role1, role2])

  await this.entityManager.save(User, [user1, user2])
}
```

写好接口后,运行一下 Nest 服务,然后用浏览器或者 postman 调一下 initData 接口,表里就会有初始化的一批数据了,其完全基于 RBAC 模式。这样,一批人员权限关系也就设计好了,如图 6.12 所示。

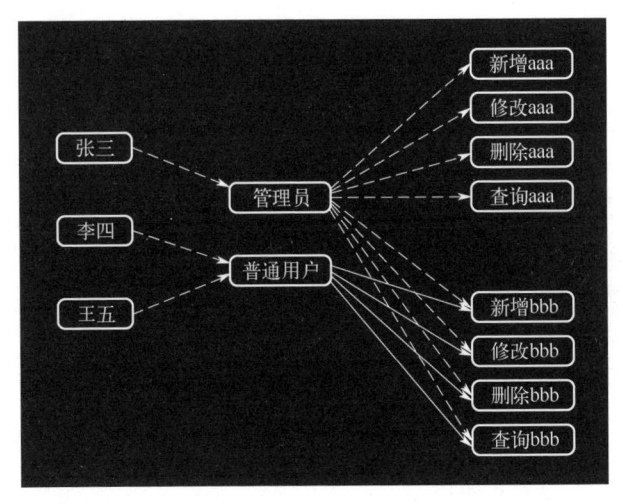

图 6.12

6.5 JWT 登录及伪造请求解决方案

有了最基本的权限设计，接下来就可以进行注册、登录、JWT 拦截校验相关的系统业务开发了。本节主要介绍 JWT 登录及伪造请求的解决方案。

JWT（JSON Web Tokens）是一种开放标准（RFC 7519），用于在各方之间安全地传输信息作为 JSON 对象。由于其信息可以被验证和信息，JWT 通常用于身份验证和信息交换。由于 JWT 较小的尺寸，它特别适合用于空间受限的环境，例如 HTTP 头部。JWT 主要由三个部分组成：

● Header（头部）：Header 通常由两部分组成，令牌的类型（即 JWT）以及所使用的签名算法（如 HMAC SHA256 或 RSA）；

● Payload（负载）：其中包含所谓的 Claims（声明），Claims 是有关实体（通常是用户）和其他数据的声明。有三种类型的 Claims：注册的声明、公共的声明和私有的声明；

● Signature（签名）：为了获取这部分的签名，必须有编码过的 Header（头部）、编码过的 Payload（负载）、一个密钥，通过 Header 中指定的算法进行签名。签名用于验证消息在传输途中有没有被更改，并且，对于使用私有密钥签署的 token，还可以验证请求方是否是 token 的合法发布者。

JWT 的使用流程一般如下：

① 用户使用各种方法（如用户名和密码）向认证服务器进行身份验证；

② 一旦服务器验证了用户的身份，它将生成一个 JWT，并将其作为响应返回给用户；

③ 用户随后将此令牌用作每次向服务器请求资源时的凭据，通常，这个令牌是放在 HTTP 请求的 Authorization 头部中，带上"Bearer"前缀；

④ 服务器会检验这个 JWT 的 signature 来确认其有效性，然后返回请求的数据。

JWT 使用上述机制使得用户状态无需在服务器持久存储，从而更容易实现无状态、可扩展的应用，这在分布式微服务架构中特别有用。同时，JWT 还提供了一种简单的方式，能够保证数据传输过程中的数据安全性和完整性。

JWT 从客户端到服务端的整个交互过程如图 6.13 所示。

JWT 身份鉴权方案中，token 会作为主要的鉴权方式来作为前后端通信校验的凭证，当该 token 被篡改或者直接被第三方得到，就可以伪造用户请求进行一系列的业务操作，是一种非常严重的安全漏洞。

有没有什么好的防止 token 被伪造的方案？对于安全防御问题，核心策略就是

"提升破解难度"。我们可以在客户端和服务端通信中加入一层额外的约定签名，签名可以由当前时间戳信息、设备 ID、日期、双方约定好的密钥等经过一些加密算法构造而成，相互约定好加密解密方式。这样攻击者就算得到了 token，还需要知道签名的生成规则和加密方式，才可以达成攻击。我们用一张图来描述一下这个增强过程，如图 6.14 所示。

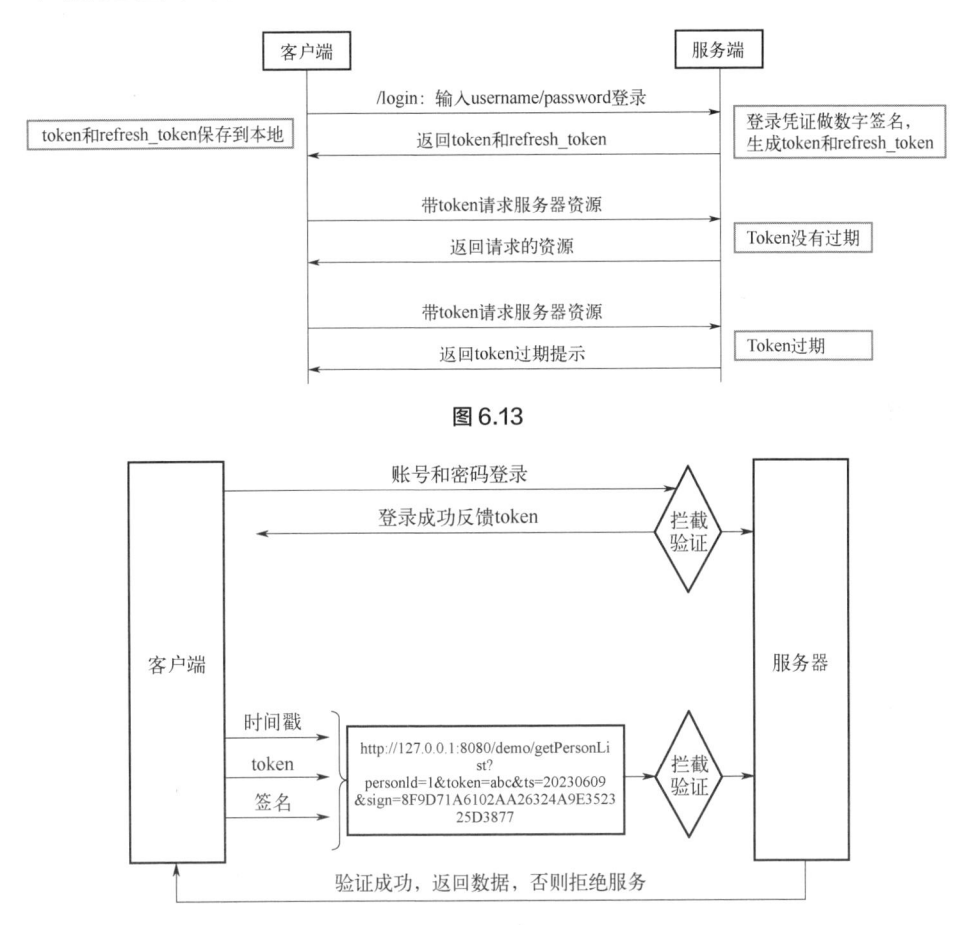

图 6.13

图 6.14

该方法其实就相当于多了一层校验判断逻辑，那么接下来我们实操一下。
① 新建一个 Nest 项目，执行命令：

```
nest new jwt-project --strict
```

② 要进行身份验证，就要准备好两个接口：
- login：登录接口，不执行 JWT 校验，用户信息通过则服务端下发 token；
- getUserInfo：业务接口，依赖于登录态，用于本案例测试。

代码如下：

```
interface loginDTO {
  access_token: string;
  msg: string;
}

interface getUserInfoDTO {
  name: string;
  id: number;
}

@Controller()
export class AppController {
  constructor(
    private readonly appService: AppService,
  ) {}

  @Post('login')
  login(): loginDTO {
    return this.appService.login();
  }

  @Post('getUserInfo')
  getUserInfo(): getUserInfoDTO {
    return this.appService.getUserInfo();
  }
}
```

③ 将项目启动，并使接口调用正常工作。接着，需要实现登录逻辑。在数据库部分，我们暂不进行配置，而是在 Nest 中维护一个数组来存储用户信息。这样做主要是为了讲解 JWT 的使用。引入 JWT 服务后的接口代码如下：

```
import { Injectable } from '@nestjs/common';
import { JwtService } from '@nestjs/jwt';

const userList = [
  {
    username: 'aaa',
    password: '123',
  },
  {
    username: 'bbb',
```

```
    password: '222',
  },
];

interface loginDTO {
  access_token: string;
  msg: string;
}

interface getUserInfoDTO {
  username: string;
  password: string;
}

@Injectable()
export class AppService {
  constructor(private readonly JwtService: JwtService) {}

  login(): loginDTO {
    // login 业务逻辑
    const user = userList[0];
    const payload = { username: user.username };
    return {
      access_token: this.JwtService.sign(payload),
      msg: '登录成功',
    };
  }

  getUserInfo(): getUserInfoDTO {
    return userList[0];
  }
}
```

④ 准备一下本案例的依赖包，安装一下：

```
npm i @nestjs/jwt
```

⑤ 在 app.module.ts 全局模块中引入 JWT 相关服务，代码如下：

```
import { Module, MiddlewareConsumer, RequestMethod } from '@nestjs/common';
import { AppController } from './app.controller';
import { AppService } from './app.service';
import { JwtModule } from '@nestjs/jwt';
import { AuthMiddleware } from './middlewares/auth.middleware';
```

```
const jwtConstants = {
  secret: 'huanchengSecret', // 密钥
  expiresIn: '3600s', // token 有效时间
};
@Module({
  imports: [
    JwtModule.register({
      secret: jwtConstants.secret,
      signOptions: { expiresIn: jwtConstants.expiresIn },
    }),
  ],
  controllers: [AppController],
  providers: [AppService],
})
```

这样调用 login 接口，验证用户信息，通过后就会基于用户的信息生成一个 JWT 字符串密钥并返回。我们在 postman 中调用一次接口试一下，如图 6.15 所示。

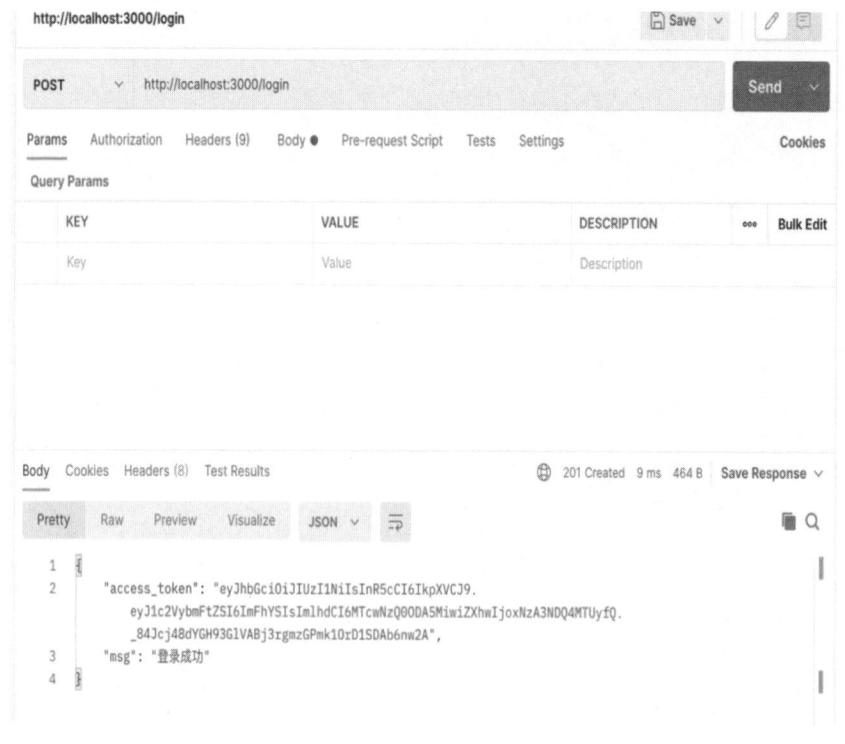

图 6.15

有了 token 就可以验证完整的客户端到服务端的会话流程了。在客户端，通常

我们会把 token 存在本地存储中。我们创建一个 React 前端项目来联调后端，后端项目先开启一下 CORD，支持客户端跨域访问。在 main.ts 中改造一下代码：

```
async function bootstrap() {
  const app = await NestFactory.create(AppModule);
  app.enableCors();
  await app.listen(3000);
}
bootstrap();
```

然后新建一个前端项目，用之前学习过的 umi.js 作为前端主框架，执行创建命令：

```
npx create-umi@latest
```

找一个页面文件，写入调用 login 的代码，如图 6.16 所示。

```
const TestPage = () => {
  useEffect(() => {
    request();
  }, []);

  const request = () => {
    fetch("http://localhost:3000/login", {
      method: "POST",
      headers: {
        "Content-Type": "application/json",
      },
    }).then((res) =>
      res.json().then((data) => {
        localStorage.setItem("access_token", data.access_token);
      })
    );
  };
```

图 6.16

为了方便测试，页面刷新一次我们就调用一次 login，更新一次 token，并且把它存在本地存储中。接下来我们准备一下 getUserInfo 相关的业务接口，对 token 加一层校验判断，代码如下：

```
@Post('getUserInfo')
getUserInfo(@Req() req: Request): getUserInfoDTO {
  try {
    const token = req.headers['authorization'];
    if (!token) {
      throw new UnauthorizedException('登录态失效,请重新登录');
    }
    this.jwtService.verify(token);
    return this.appService.getUserInfo();
```

第 6 章 全栈化与 Serverless 229

```
  } catch (error) {
    throw new UnauthorizedException(
      error.message || 'Unauthorized: Invalid token',
    );
  }
}
```

对于 token 作为空和真实性的两层校验，做了一次增强，此时在前端删除 token，再请求一次接口，请求头中没有 authorization 字段，如图 6.17 所示。

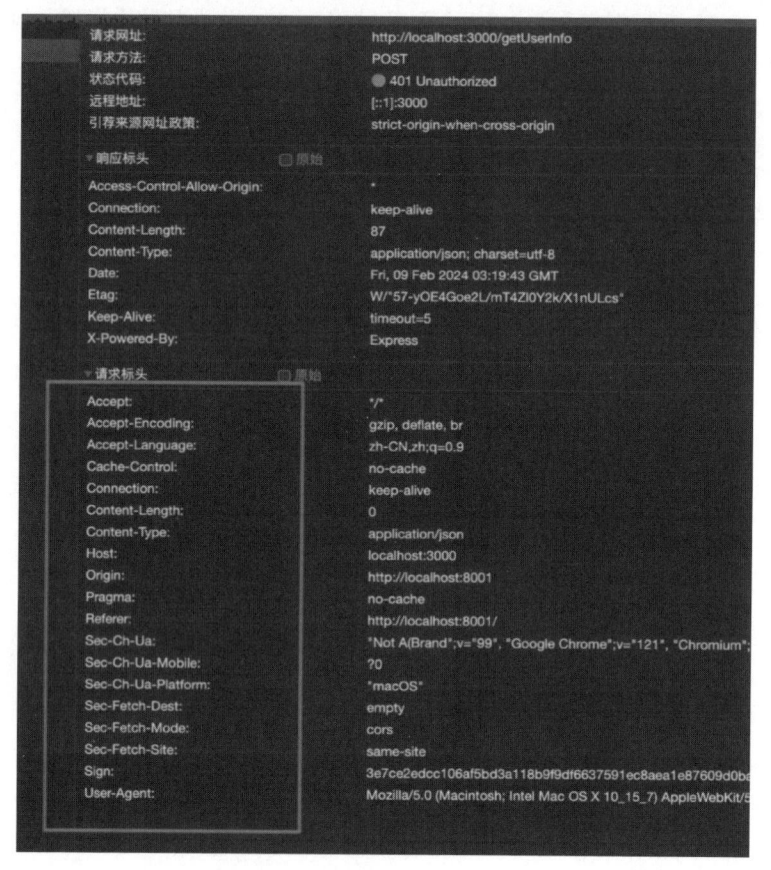

图 6.17

后端报了 401 的异常，说明运行到我们前面所加的 token 为空的判断中了，符合预期。接着修改一下 token，再调用一次，抛出了 JWT 校验失败的异常，如图 6.18 所示。

至此，基于 JWT 的登录身份验证部分已经实现了。但当前我们的 token 是暴露在浏览器端的，如果复制了这个 token，去 postman 这种无状态的端调用一下，是否也可以调通呢？为了验证这个猜想，我们测试一下，如图 6.19 所示，在 postman 端利用浏览器中的 token 仍可调用服务端的接口。

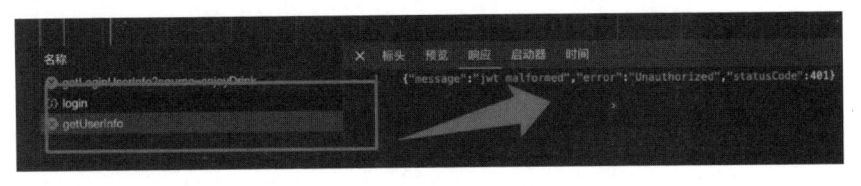

图 6.18

图 6.19

如何限制类似的伪造请求呢？本节开头也提到了，加一层安全认证即可，在传递 token 的鉴权接口处我们再多加一个 sign 验签字段，并且也存放在 Header 中。这里的加密解密我们用 crypto-js，先改造一下前端代码，安装一下依赖包：

```
npm i crypto-js --save-dev
```

然后在测试页面中引入 crypto-jsku 库。我们采用客户端加密 sign、服务端解密 sign 的方案，因此，客户端实现加密的代码如下：

```
//十六位十六进制数作为密钥
const SECRET_KEY = CryptoJS.enc.Utf8.parse('3333e6e143439161');
//十六位十六进制数作为密钥偏移量
const SECRET_IV = CryptoJS.enc.Utf8.parse('e3bbe7e3ba84431a');
const encrypt = (data: object | string): string => {
  //加密
  if (typeof data === 'object') {
    try {
```

第 6 章 全栈化与 Serverless 231

```
    data = JSON.stringify(data);
  } catch (e) {
    throw new Error('encrypt error' + e);
  }
}
const dataHex = CryptoJS.enc.Utf8.parse(data);
const encrypted = CryptoJS.AES.encrypt(dataHex, SECRET_KEY, {
  iv: SECRET_IV,
  mode: CryptoJS.mode.CBC,
  padding: CryptoJS.pad.Pkcs7,
});
return encrypted.ciphertext.toString();
};
```

加密需要密钥和密钥偏移量，这两个参数在前端和后端约定好保持一致即可，这点非常重要。然后我们基于上文实现的加密方法，在 getUserInfo 接口的请求头中把密钥携带上去，如图 6.20 所示。

```
const token = localStorage.getItem("access_token");
const sign = encrypt(
  new Date().getTime() + ":" + "jack" + ":" + token?.slice(0, 6)
);
const headers: any = {
  "Content-Type": "application/json",
  sign,
};
```

图 6.20

加密的基准字符串由调用时间、用户名、token 前 6 位组成，中间用冒号隔开，就像这样：

```
"1707449642869:jack:eyJhbG"
```

如此，前端代码就改造好了，然后我们改造一下后端代码。同样也是先把依赖包安装好，由于通用的是 JavaScript 包，所以安装方式一样。然后在 Nest 中实现一下解密验证的逻辑即可，代码如下：

```
@Controller()
export class AppController {
  constructor(
    private readonly appService: AppService,
    private readonly jwtService: JwtService,
  ) {}

  private readonly SECRET_KEY = CryptoJS.enc.Utf8.parse
```

```typescript
('3333e6e143439161');
private readonly SECRET_IV = CryptoJS.enc.Utf8.parse
('e3bbe7e3ba84431a');

private decrypt(data: string): string {
  //解密方法
  const encryptedHexStr = CryptoJS.enc.Hex.parse(data);
  const str = CryptoJS.enc.Base64.stringify(encryptedHexStr);
  const decrypt = CryptoJS.AES.decrypt(str, this.SECRET_KEY, {
    iv: this.SECRET_IV,
    mode: CryptoJS.mode.CBC,
    padding: CryptoJS.pad.Pkcs7,
  });
  const decryptedStr = decrypt.toString(CryptoJS.enc.Utf8);
  return decryptedStr.toString();
}

@Post('login')
login(): loginDTO {
  return this.appService.login();
}

@Post('getUserInfo')
getUserInfo(@Req() req: Request): getUserInfoDTO {
  try {
    const token = req.headers['authorization'];
    const sign = req.headers['sign'];
    if (!token) {
      throw new UnauthorizedException('登录态失效，请重新登录');
    }
      if (!sign) {
        throw new UnauthorizedException('身份验证失败');
      }
      this.jwtService.verify(token);
      const signRes = this.decrypt(sign as string);
      const [time, userName, tokenStr] = signRes.split(':');
      console.log('time:', time);
      console.log('userName:', userName);
      console.log('tokenStr:', tokenStr);
      if (userName === '' || time === '' || tokenStr === '') {
        throw new UnauthorizedException('身份验证失败');
      }
```

```
    if (userName !== 'aaa') {
      throw new UnauthorizedException('用户信息错误');
    }
    return this.appService.getUserInfo();
  } catch (error) {
    throw new UnauthorizedException(
      error.message || 'Unauthorized: Invalid token',
    );
  }
}
```

上述代码中，我们在后端加上了解密的代码，同样基于密钥、密钥偏移量，与客户端保持一致。我们在业务接口中 token 验证的基础上增加了一层对于 sign 的验证，加强了接口安全性，接下来再测试一下。首先，只传 token 不传 sign 的方案失败了，后端抛出了身份验证失败。本节中，后端默认是使用第一个 user 来返回的，sign 基于用户信息来组成，那前端传一个不一样的 username 作为 sign，加上一层 sign 定制判断逻辑，代码如下：

```
if (userName !== 'aaa') {
  throw new UnauthorizedException('用户信息错误');
}
```

在前端请求一次该接口进行测试，返回了用户信息错误的异常，也命中了接口中错误捕获的逻辑，然后我们看一下后端打印出来的解密 sign 的日志结果，如图 6.21 所示。

图 6.21

打印出来的用户名是 jack，而不是 aaa，命中了校验失败规则。如此，我们对伪造请求的校验方案也实现了。

当我们需要类似逻辑时，可以考虑把该方案封装到一个中间件上去（Middleware），即新增一个 middlewares/auth.middleware.ts，代码如下：

```typescript
import {
  Injectable,
  NestMiddleware,
  UnauthorizedException,
} from '@nestjs/common';
import { Request, Response, NextFunction } from 'express';
import { JwtService } from '@nestjs/jwt';
import * as CryptoJS from 'crypto-js';

@Injectable()
export class AuthMiddleware implements NestMiddleware {
  constructor(private readonly jwtService: JwtService) {}

  private readonly SECRET_KEY = CryptoJS.enc.Utf8.parse('3333e6e143439161');
  private readonly SECRET_IV = CryptoJS.enc.Utf8.parse('e3bbe7e3ba84431a');

  private decrypt(data: string): string {
    //解密
    const encryptedHexStr = CryptoJS.enc.Hex.parse(data);
    const str = CryptoJS.enc.Base64.stringify(encryptedHexStr);
    const decrypt = CryptoJS.AES.decrypt(str, this.SECRET_KEY, {
      iv: this.SECRET_IV,
      mode: CryptoJS.mode.CBC,
      padding: CryptoJS.pad.Pkcs7,
    });
    const decryptedStr = decrypt.toString(CryptoJS.enc.Utf8);
    return decryptedStr.toString();
  }
  use(req: Request, res: Response, next: NextFunction) {
    try {
      const token = req.headers['authorization'];
      const sign = req.headers['sign'];
      if (!token) {
        throw new UnauthorizedException('登录态失效，请重新登录');
      }
      if (!sign) {
```

```
      throw new UnauthorizedException('身份验证失败');
    }
    // token 校验
    this.jwtService.verify(token);
    // sign 校验
    const signRes = this.decrypt(sign as string);
    const [time, userName, tokenStr] = signRes.split(':');
    console.log('time:', time);
    console.log('userName:', userName);
    console.log('tokenStr:', tokenStr);
    if (userName === '' || time === '' || tokenStr === '') {
      throw new UnauthorizedException('身份验证失败');
    }
    if (userName !== 'aaa') {
      throw new UnauthorizedException('用户信息错误');
    }
    return next();
  } catch (error) {
    throw new UnauthorizedException(
      error.message || 'Unauthorized: Invalid token',
    );
  }
}
```

然后在 Nest 项目中开启这个中间件。目前的 login 接口不需要校验，因此只开启 getUserInfo 接口即可，代码如下：

```
export class AppModule {
  configure(consumer: MiddlewareConsumer) {
    consumer
      .apply(AuthMiddleware)
      .exclude(
        { path: '/login', method: RequestMethod.POST },
      )
      .forRoutes({ path: '*', method: RequestMethod.ALL });
  }
}
```

简单总结一下，JWT 是目前主流的鉴权方案，但是在不执行安全方案的情况下，直接把 token 暴露在请求体中很容易引发伪造请求的问题。通过在请求中多加一层加密解密的约定式验签流程，可以在通信中增加安全性。

6.6 跨端扫码登录

扫码登录是一种便捷的身份验证方式,用户通过扫描屏幕上显示的二维码(QR码)来登录应用或网站。这种登录方式在移动设备普及的情况下变得越来越流行,因为它提供了一个无需记忆密码的安全登录方案。比如微信,当用户需要在 PC 上登录微信时,PC 端就会展示一个二维码,用户打开手机端微信扫码,进入一个页面,确认登录即可。

这中间的过程其实很简单,PC 端轮询二维码的状态,H5 端更改二维码状态并在点击"确认登录"动作的时候将已有的登录态,也就是 token 传给服务端,服务端进行一层校验,校验通过就把二维码状态更改,基于用户信息返回给正在轮询的 PC 端一个 token 即可。

扫码登录的系统工作过程如图 6.22 所示。

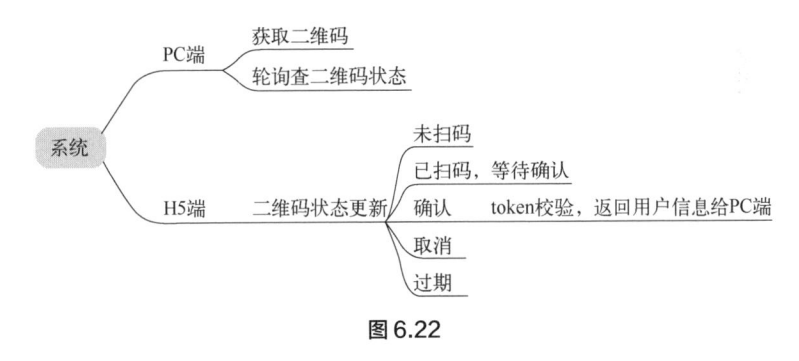

图 6.22

搞清楚扫码登录实现原理以后,我们接下来就开始用 Nest.js 来实现。
① 新建一个 Nest 项目,执行命令:

```
nest new qrcode-project
```

新建到运行项目服务启动的过程如图 6.23 所示。
② 通过 nest cli 命令创建出一个通用的扫码模块服务,执行以下命令:

```
nest n mo qrcode
nest n s qrcode
nest n co qrcode
```

③ 安装一下生成二维码所需要的依赖包,在项目根目录执行命令:

```
npm i qrcode --save-dev
```

④ 开始业务部分的开发,基于图 6.20 中的思路图,其实总共只需要三个接口。对于 PC 端,需要准备创建二维码、查询二维码状态这两个动作;对于 H5 端,只

需要在进行了某一步操作后，把二维码状态更新到服务端再同步给 PC 端即可。

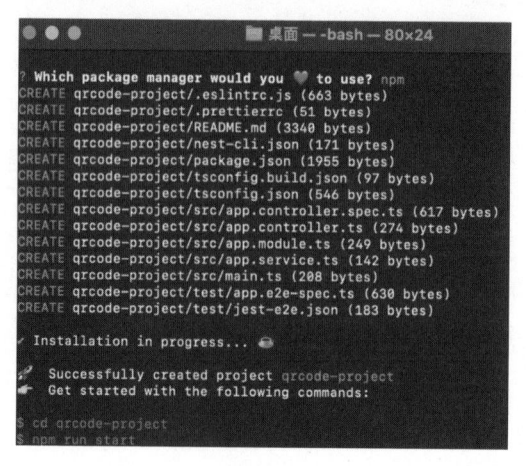

图 6.23

- create：创建二维码，提供给 PC 端扫码；
- getQrcodeStatus：获取二维码状态；
- updateQrcodeStatus：更新二维码状态。

我们来把这三个接口实现一下，代码如下：

```
import { BadRequestException, Controller, Post, Get } from '@nestjs/common';
import { randomUUID } from 'crypto';
import * as qrcode from 'qrcode';

const map = new Map<string, QrCodeInfo>();

type QrCodeStatus =
  | 'noscan' // 未扫码
  | 'scan-wait-confirm' // 已扫码，等待确认
  | 'scan-confirm' // 确认
  | 'scan-cancel' // 取消
  | 'expired'; // 过期
interface QrCodeInfo {
  status: QrCodeStatus;
  userInfo?: {
    userId: number;
  };
}
```

```
/**
 * @description: PC 端获取二维码
 * @return {*}
 */
@Controller('qrcode')
export class QrcodeController {
  @Get('/create')
  async create() {
    const url = 'http://localhost:8000/h5Login';
    const qrcodeId = randomUUID();
    const qrcodeUrl = await qrcode.toDataURL(url);
    map.set(qrcodeId, {
      status: 'noscan',
    });
    return {
      qrcodeId,
      qrcodeUrl,
    };
  }

  /**
   * @description: PC 端轮询获取
   * @return {*}
   */
  @Get('/getQrcodeStatus')
  async getQrcodeStatus(qrcodeId: string) {
    const info = map.get(qrcodeId);
    if (info.status === 'scan-confirm') {
      return info.userInfo;
    }
    return {
      status: info.status,
    };
  }

  /**
   * @description: H5 端修改二维码状态（扫码、确认登录）
   * @return {*}
   */
  @Post('/updateQrcodeStatus')
  async updateQrcodeStatus(qrcodeId: string, status: QrCodeStatus){
    const info = map.get(qrcodeId);
```

```
      if (!info) {
        throw new BadRequestException('二维码已过期');
      }
      info.status = status;
      return {
        status: info.status,
      };
    }
  }
```

梳理一下三个接口的逻辑，/qrcode/create 会返回一个二维码图片地址和二维码的 ID，PC 端展示出来图片即可。二维码的链接就是 H5 授权登录的页面，可以选择把二维码 ID 带到 url query 里去查询。

然后 PC 端就开始执行轮询任务，每次带着这个二维码 ID 去查询即可，可以进行一些更新的交互动作。

对于/qrcode/updateQrcodeStatus，在 H5 端如果点击"取消"或者"确认"，就传不同的二维码 status 枚举给服务端同步即可。

当点击"登录"后，二维码更新到确认的状态后，还需要返回用户信息给 PC 端，也就是返回 token，那这里就需要做登录鉴权动作。

接下来我们安装 JWT 包，执行命令：

```
npm install @nestjs/jwt
```

然后在 App 模块中导入全局注入 JWT 模块，配置方式和 6.5 小节所述一样。接着我们就需要改造/qrcode/updateQrcodeStatus、/qrcode/getQrcodeStatus 接口，当 H5 端确认登录时，服务端就需要先进行一层 token 验证，如果验证通过，就把用户信息保存在二维码表中，就像这样：

```
@Post('/updateQrcodeStatus')
async updateQrcodeStatus(
  qrcodeId: string,
  status: QrCodeStatus,
  @Headers('Authorization') auth: string,
) {
  const info = map.get(qrcodeId);
  if (!info) {
    throw new BadRequestException('二维码已过期');
  }
  if (status === 'scan-confirm' && auth) {
    // 确认登录
    try {
```

```
      const [, token] = auth.split(' ');
      const info = await this.jwtService.verify(token);
      const user = users.find((item) => item.id == info.userId);
      info.userInfo = user;
      return user;
    } catch (e) {
      throw new UnauthorizedException('token 过期，请重新登录');
    }
  }
  info.status = status;
  return {
    status: info.status,
  };
}
```

如果二维码更新成功了，PC 端下一次轮询就可以得到确认的二维码状态，并且可以得到 userInfo，再基于用户信息返回一个 token 给 PC 端即可，这样就登录成功了。/qrcode/getQrcodeStatus 接口代码改造如下：

```
@Get('/getQrcodeStatus')
async getQrcodeStatus(qrcodeId: string) {
  const info = map.get(qrcodeId);
  if (info.status === 'scan-confirm') {
    const accsess_token = this.jwtService.sign(info.userInfo);
    return { userInfo: info.userInfo, accsess_token };
  }
  return {
    status: info.status,
  };
}
```

扫码登录到这里就实现了，其原理其实很简单，只是涉及了跨端，很多人会觉得这很复杂。因此我们做一个总结，扫码登录的主要思路可以概括为以下几个步骤：

① 二维码的展示：当用户想要登录时（通常在 Web 应用或者 PC 软件中），服务器会生成一个包含特定登录信息的二维码并展示给用户。

② 二维码的扫描：用户使用手机上的应用（例如移动端的 App 内置扫码功能或通用的二维码扫描应用）扫描二维码，扫码功能得以识别出二维码内包含的登录信息。

③ 登录验证：移动应用识别二维码的信息后（一般包含认证服务器的 URL 和一段用于两端验证的 token），移动应用通过 https 协议将 token 发送到服务器进

行验证。

④ 会话创建：服务器验证 token 的有效性，若验证通过，则会在服务器端创建一个新的登录会话，并返回会话信息给 Web 应用或 PC 软件。

⑤ 登录确认：桌面端或 Web 端接收到服务器响应的登录会话之后，更新用户界面，确认用户已经登录成功。

6.7　Midway.js 入门

从本节开始，将介绍另一款全栈框架——Midway.js。了解 Midway.js，开发者在技术选型中就有了更多的选择。

Midway Serverless 是用于构建 Node.js 云函数的 Serverless 框架，可以帮助开发者在云原生时代大幅降低维护成本，更专注于产品研发。其专注于函数即服务，开发者只需要编写 JavaScript 函数即可，像编写 Java 接口一样简单，并且其提供了开箱即用的部署解决方案。

Midway 是一个适用于 Node.js 的全栈框架，它建立在著名的 Express 和 Koa 框架之上，致力于提供更加易用的服务端开发体验。Midway 通过支持 TypeScript，具有一系列现代软件工程的特性，包括面向对象编程、依赖注入（DI）和装饰器（Decorator）等，极大地增强了 Node.js 应用的可维护性和可扩展性。

Midway 主要特点有：

- 强类型语言：使用 TypeScript 作为开发语言，提供了静态类型检查和最新的 ECMAScript 特性；
- 依赖注入：内置 IOC 容器，通过装饰器等方式实现依赖注入，减少模块间耦合；
- 框架无关性：可在多种常见 Node.js 框架，如 Express、Koa 中使用，或用于搭建自定义框架；
- 丰富的生态系统：与 Egg.js 高度兼容，方便从 Egg.js 中迁移，并能利用 Egg.js 丰富的插件生态；
- 面向切面编程：支持面向切面编程（AOP），方便开发者组织、复用代码；
- 适合大型应用：适合构建大型企业级应用，有助于提高开发、测试、部署的效率；
- 云原生支持：内置对 Serverless 应用的支持，可以轻松部署到云函数环境。

Midway 和 Nest 有许多的相似之处：

- 都全面支持 TypeScript；
- 都提供了装饰器和依赖注入特性，以改善代码组织；

- 都设计成了模块化，易于扩展和维护。

但二者也有一些差异：

- 生态兼容性：Midway 更侧重于 Egg.js 的生态，非常适合已有基于 Egg.js 开发的应用的团队，而 Nest 直接构建自己的生态；
- 架构影响：Nest 受到了 Angular 的很大影响，适用于熟悉 Angular 的前端开发者，而 Midway 则试图将类似的概念引入了后端；
- Serverless 支持：Midway 更专注于对 Serverless 的内置支持，而 Nest 则需要额外的插件和适配层；
- 框架定位：Nest 更专注于后端 API 的构建，Midway 更专注于成为一个多功能的全栈框架。

选择哪个框架取决于项目需求、开发团队的背景和偏好。如果团队已经熟悉 Express/Koa 或 Egg.js，并且考虑到将来可能的 Serverless 部署，Midway 可能是一个更适合的选择。如果团队更喜欢 Angular 的架构理念，并且专注于构建复杂的后端应用和微服务，Nest.js 可能更合适。

Midway.js 提供了面向对象（OOP+Class+Ioc）和函数式（FP+Function+Hook）的开发范式。面向对象写法，采用了类+装饰器的形式，就像这样：

```ts
// src/controller/home.ts
import { Controller, Get } from '@midwayjs/core';
import { Context } from '@midwayjs/koa';

@Controller('/')
export class HomeController {
  @Inject()
  ctx: Context

  @Get('/')
  async home() {
    return {
      message: 'Hello Midwayjs!',
      query: this.ctx.ip
    }
  }
}
```

对于函数式，提供了和 React 很相像的一种写法，就像这样：

```ts
import { useContext } from '@midwayjs/hooks'
import { Context } from '@midwayjs/koa';
```

```
export default async function home () {
  const ctx = useContext<Context>()
  return {
    message: 'Hello Midwayjs!',
    query: ctx.ip
  }
}
```

接下来我们将以 Oop + Class + Ioc 的范式来进行入门介绍演示。首先我们初始化一个 Midway 项目，只需要两行命令，即可启动一个后端服务：

```
npm init midway
npm run dev
```

在 controller 目录中，新建一个 src/controller/weather.controller.ts 文件，内容如下：

```
import { Controller, Get } from '@midwayjs/core';

@Controller('/')
export class WeatherController {
  // 这里是装饰器，定义一个路由
  @Get('/weather')
  async getWeatherInfo(): Promise<string> {
    // 这里是 http 的返回，可以直接返回字符串，数字，JSON，Buffer 等
    return 'Hello Weather!';
  }
}
```

这就是控制器，概念和 Nest 是一样的，在后端直接注册一个接口。这样就可以通过前端请求的形式获取/weather 接口了，就像这样：

```
fetch('http://127.0.0.1/weather').then(res => {
  res.json().then(data => {
    console.log(data); // Hello Weather
  })
})
```

对于@Controller，可以理解为一个后端项目通过一个控制器来启动一个接口，在里面包含了许多模块的服务，如 user 类、list 类、upload 类等。而 user 中可能包含了注册、登录、注销；upload 中可能包含了上传、删除图片等，因此其接口看起来会像是这样的：

```
import { Controller, Post, Inject, Query, Get } from '@midwayjs/core';
import { UserService } from './service/user.service';
import { ListService } from './service/list.service';
```

```
@Controller('/')
export class CommonController {
  @Inject()
  ctx;

  @Inject()
  UserService: UserService;

  @Inject()
  ListService: ListService;

  @Inject()
  UploadService: UploadService;

  @Post('/register')
  async register(@Query('userId') userId: string, @Query('password')
  password: string): Promise<boolean> {
    return this.UserService.register({userId, password});
  }
  @Get('/getUserInfo')
  async getUserInfo() {
    return this.UserService.getUserInfo();
  }
  // List 和 Upload 的接口...
}
```

在 user.service.ts 中，与 Nest.js 一样，涵盖了接口中的所有业务逻辑部分，controller 只负责转发和注册。服务部分的代码如下：

```
import { Provide, Inject, Context } from '@midwayjs/core';

interface UserInfo {
  userName: string;
  age: number;
  sex: string;
}

@Provide()
export class UserService {
  @Inject()
  ctx: Context;
  async register(params): Promise<boolean> {
```

```
    // 注册逻辑
    return true;
  }

  async getUserInfo(): Promise<UserInfo> {
    // 获取用户信息逻辑
    return {
      userName: '量子前端',
      age: 20,
      sex: '不详'
    };
  }
}
```

有没有感觉到编写一个接口就像是写一个函数或者是类一样简单呢？Midway 还提供了很多强大的功能，如中间件、组件、Http 服务等，这部分不再赘述，具体可以去看官方文档。我们直接来实践两个全栈场景，分别是图片上传和验证码，来具体感受一下。

首先是图片上传，我们直接下载 Midway 官方提供的 upload 依赖包，在项目根目录中执行命令：

```
npm i @midwayjs/upload@3 --save
```

然后在 configuration.ts 中导入 upload 依赖包，对应了 Nest 中的 app.module.ts 全局模块。值得一提的是，Midway 并没有进行模块化，更像是 umi.js。注册服务代码如下：

```
@Configuration({
  imports: [upload],
  importConfigs: [
    {
      default: defaultConfig,
      prod: prodConfig,
    },
  ],
  conflictCheck: true,
})
```

接下来我们在控制器中去使用它，和开发后端一样，直接准备 upload 的控制器、服务，控制器部分的代码如下：

```
import { Controller, Post, Inject, Files } from '@midwayjs/core';
import { UploadService } from './service/upload.service';
```

```
@Controller('/')
export class CommonController {
  @Inject()
  UploadService: UploadService;

  @Post('/upload')
  async upload(@Files() files): Promise<string[]> {
    return this.UploadService.upload(files);
  }
}
```

在 controller 层不进行外调服务，这是概念性原则。直接在 upload.service.ts 中注入刚才安装的外部 upload 服务来编写业务逻辑，代码如下：

```
import { Provide, Inject, Context } from '@midwayjs/core';
import * as path from 'path';
import * as moment from 'moment';
import * as uuid from 'uuid';
import * as fs from 'fs';

@Provide()
export class UploadService {
  @Inject()
  ctx: Context;

  async upload(files): Promise<string[]> {
    const fileDir = path.join(this.ctx.app.getBaseDir(), '..', 'public');
    const timeDir =
     `${moment().format('YYYY')}/${moment().format('MM-DD')}`;
    const url = path.join(fileDir, timeDir);
    const fileList = [];
      if (!fs.existsSync(url)) fs.mkdirSync(url, { recursive: true });
    for (let i = 0; i < files.length; i++) {
      const file = files[i];
        const extname: string = path.extname(file.filename).toLowerCase();
      const data = fs.readFileSync(file.data);
      const fileName = uuid.v1();
        const target = path.join(url, `${fileName}${extname}`);
      fs.writeFileSync(target, data);
      fileList.push(`${url}/${fileName}${extname}`);
    }
```

```
    return fileList;
  }
}
```

上述代码中，接口接收前端传来的 file 类型数组，后端解析文件，并且通过 uuid+moment 的形式给文件提供命名，然后写入后端的文件夹中。官方提供了文件上传和流上传两种模式，这里以文件上传的方式将图片保存在 Midway 项目的 public 目录中。

接下来我们可以写一个前端请求来测试一下：

```
const fileUpload = (e) => {
  const formData = new FormData();
  formData.append('file', e.target.files[0]);
  console.log(e.target.files);
  fetch('http://127.0.0.1:7002/upload', {
    method: 'POST',
    body: formData,
  }).then((res) => {
    res.json().then((data) => {
      // 获取到图片上传的 fileList，回显在 DOM 中
    });
  });
}

return (
  <input type="file" onChange={fileUpload} />
)
```

这样，一个简单基础版本的图片上传接口就完成了。接下来，通过一个验证码校验的实例进行验证。

首先还是安装官方提供的服务依赖包，在项目根目录执行命令：

```
npm i @midwayjs/captcha@3 --save
```

然后在 configuration.ts 中导入这个服务依赖包，全局开启它：

```
@Configuration({
  imports: [captcha],
  importConfigs: [
    {
      default: defaultConfig,
      prod: prodConfig,
    },
```

```
  ],
  conflictCheck: true,
})
```

创建一个验证码的控制器和服务，声明两个接口，分别是获取验证码接口和校验验证码接口以业务中最常见的图形验证码为例：

```
import { Controller, Post, Inject, Get } from '@midwayjs/core';
import { CaptchaService } from '@midwayjs/captcha';

@Controller('/')
export class CommonController {
  @Inject()
  CaptchaService: CaptchaService;

  @Get('/get-image-captcha')
  async getImageCaptcha() {
      const { id, imageBase64 } = await this.CaptchaService.image({
        width: 120,
        height: 40,
        size: 6,
        type: 'number',
      });
      return {
        id, //验证码 ID
        imageBase64, //验证码 SVG 图片的 base64 数据，可以直接放入前端的 img 标
          签内
      };
  }

  //校验验证码是否正确
  @Post('/check-captcha')
  async getCaptcha() {
    const { id, answer } = this.ctx.request.body;
      const passed: boolean = await this.CaptchaService.check(id, answer);
    return passed;
  }
}
```

这里我们直接调用了官方的服务接口：

- 获取验证码接口直接返回给了前端一个验证码图片 ID 和图片 base64 地址用于展示；

- 校验验证码接口将前端验证的结果和获取的验证码 ID 传给后端来校验是否一致。

这里简单写一段 React 弹窗组件的代码来测试一下:

```
const openCheckCaptchaModal = () => {
  // 获取所有 tab 的商户数量
  fetch('http://127.0.0.1:7002/get-image-captcha').then((res) => {
    res.json().then(({ id, imageBase64 }) => {
      Modal.alert({
        content: (
          <>
            <img src={imageBase64} />
            <Form form={form}>
              <Form.Item name="captcha">
                <Input placeholder="请输入验证码" />
              </Form.Item>
            </Form>
            <span
              onClick={() => {
                Modal.clear();
                openCheckCaptchaModal();
              }}
            >
              换一张
            </span>
          </>
        ),
        onConfirm: () => {
          const captcha = form.getFieldValue('captcha');
          if (captcha) {
            fetch('http://127.0.0.1:7002/check-captcha', {
              method: 'POST',
              body: JSON.stringify({
                id,
                answer: captcha,
              }),
              headers: {
                'Content-Type': 'application/json',
              },
            }).then((res) => {
              res.json().then((data) => {
                if(data) {
```

```
                    return Message.success('验证成功');
                }
            });
        });
    }
        },
    });
  });
});
};
return (
  <span onClick={openCheckCaptchaModal}>验证</span>
)
```

这段代码中我们的页面很简单，就是一个验证元素加上点击事件。当点击事件触发时，会先请求获取验证码接口，得到验证码图片base64地址以后，打开一个antd弹窗组件，展示了一个表单，包含了验证码图片和一个输入框；当点击"换一张"时，将再调用一次获取验证码接口，更新图片；当点击"我知道了"按钮确认时，将会校验前端输入的验证码与实际验证码是否一致。

获取验证码的请求如图6.24所示。

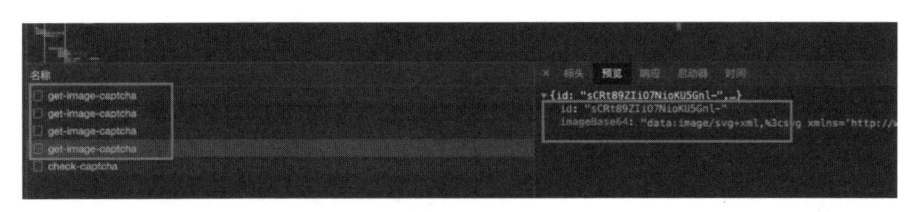

图6.24

校验验证码的请求如图6.25所示。

图6.25

这两个案例中，其实我们没有自己写服务，只是运用了官方提供的开箱即用的三方服务以及感受了一下Midway的开发模式。Midway.js与Nest.js的不同之处，在于 Midway.js 项目执行 npm run deploy 即可进入部署流程，部署非常方便，只需前置准备阿里云或者其他服务器账号即可。阿里云首次部署需要 accountId、acountKey、accountSecret。

6.8 Midway.js 实现注册、登录、鉴权

本节将从项目搭建开始开发一个实现注册、登录、鉴权的简易版注册登录系统，主要的功能和技术选型如下：

- 服务端框架：Midway.js；
- 密码加密存储：bcrypt.js；
- 数据库存储：typeorm、mysql；
- 登录鉴权：jwt。

其数据库和鉴权方案与 Nest 一样，都是通用方案，用各框架集成的包即可。我们安装一下 mysql 环境，建好数据库和一张用户 user 表，Dbeaver 或者 vscode database 都可以。本节通过后者来初始化数据，表结构如图 6.26 所示。

准备好环境后，我们创建一个 Midway 项目，执行命令：

```
npm init midway@latest -y
```

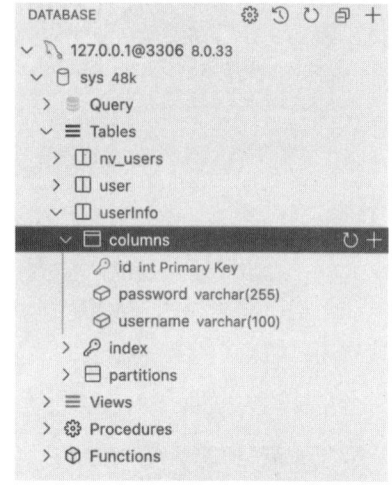

图 6.26

然后安装所有要用到的第三方依赖包，在项目根目录中执行安装命令：

```
npm i @midwayjs/typeorm@3 typeorm mysql2 -save
```

接下来在 configuration.ts 中引入 typeorm 组件：

```
// configuration.ts
import { Configuration } from '@midwayjs/core';
import * as orm from '@midwayjs/typeorm';
import { join } from 'path';

@Configuration({
  imports: [
    // ...
    orm
  ],
  importConfigs: [
    join(__dirname, './config')
```

252　深入浅出 React.js：原理与实战

```
  ]
})
```

这一点和 Nest 是很类似的, 然后在 Midway 的 config 目录中配置数据库信息:

```
import { MidwayConfig } from '@midwayjs/core';
import { User } from '../entity/user.entity';

export default {
  // use for cookie sign key, should change to your own and keep security
  keys: '1697424147281_6188',
  koa: {
    port: 7001,
  },
  typeorm: {
    dataSource: {
      default: {
        /**
         * 单数据库实例
         */
        type: 'mysql',
        host: 'localhost',
        port: 3306,
        username: 'root',
        password: 'xxxxx',
        database: '数据库名',
          synchronize: false, // 如果第一次使用, 不存在表, 有同步的需求可以写
            true, 注意会丢数据
        logging: false,
        // 配置实体模型
        entities: [User],
      },
    },
  },
} as MidwayConfig;
```

可以看到, mysql 数据库的配置项与 Nest.js 中是一样的, 因为框架层只做底层设计, 而数据库属于第三方依赖, 因此在 Nest 和 Midway 中通用的场景还是有很多的。最后我们还需要一个数据表实例。新建/entity/user.entity.ts, 配置方式与 Nest 中的实体完全一致, 因为引用的都是 typeorm 的能力, 代码如下:

```
import { Entity, Column, PrimaryColumn } from 'typeorm';
```

```
@Entity('userInfo')
export class User {
  @PrimaryColumn()
  id: number;

  @Column()
  username: string;

  @Column()
  password: string;
}
```

到这里，关于数据库的配置环境已经创建好了，项目运行起来后也会自动和数据库关联起来。接下来我们开始写业务部分的代码。首先是注册登录接口，我们新建两个文件，分别是 user.controller.ts 文件和 user.service.ts 文件，controller 用于中转服务，service 用于存放业务逻辑代码。uer.controller.ts 代码很简单，直接关联到对应的函数服务即可，没什么特殊之处，代码如下：

```
import { Inject, Controller, Post } from '@midwayjs/core';
import { Context } from '@midwayjs/koa';
import { UserService } from '../service/user.service';

@Controller('/api')
export class APIController {
  @Inject()
  ctx: Context;

  @Inject()
  userService: UserService;

  @Post('/register')
  async register() {
    const params = this.ctx.request.body as {
      username: string;
      password: string;
    };
    const user = await this.userService.register(params);
    return { success: true, message: 'OK', data: user };
  }

  @Post('/login')
  async login() {
```

```
    const params = this.ctx.request.body as {
      username: string;
      password: string;
    };
    const user = await this.userService.login(params);
    return { success: true, message: 'OK', data: user };
  }
}
```

然后把服务层的雏形创建出来，再把具体的入参和出参定义写入：

```
import { Provide, httpError, Inject, Context } from '@midwayjs/core';
import { User } from '../entity/user.entity';
import { InjectEntityModel } from '@midwayjs/typeorm';
import { Repository } from 'typeorm';
const { v4: uuidv4 } = require('uuid');
import { JwtService } from '@midwayjs/jwt';

@Provide()
export class UserService {
  @InjectEntityModel(User)
  userModal: Repository<User>;

  @Inject()
  jwtService: JwtService;

  @Inject()
  ctx: Context;

  async register(options: { username: string; password: string }) {
    const { username, password } = options;
    return {
      success: true,
      username,
      res: '注册成功',
    };
  }

  async login(options: { username: string; password: string }) {
    const { username, password } = options;
    return {
      accessToken: 'xxxx',
    };
```

```
    }
  }
```

前面的代码已经让项目和数据库关联好了，接下来需要让接口和数据表绑定起来。我们可以通过 InjectEntityModal 在接口服务中注入表信息，来进行增、删、改、查操作。有了操作数据库的能力，就可以开始开发主体逻辑了。

我们先看注册部分，主要逻辑如下：

- 传入用户信息，去数据库查重；
- 用户已存在，返回异常；
- 用户不存在，执行数据表写入语句。

代码如下：

```
async register(options: { username: string; password: string }) {
  const { username, password } = options;
  const user = new User();
  const findRes = await this.userModal.findOne({
    where: {
      username,
    },
  });
  if (findRes) return new httpError.BadRequestError('用户已存在');
  user.id = uuidv4();
  user.username = username;
  user.password = password;
  const res = await this.userModal.save(user);

  return {
    success: true,
    username,
    res: '注册成功',
  };
}
```

通过接口调用软件来模拟一个请求，如图 6.27 所示。

注册接口返回成功的响应报文了，此时我们去数据库中检查一下，发现用户信息已经被成功插入数据库中了，如图 6.28 所示。

相同的入参再调用一次，会返回重复用户的异常信息，如图 6.29 所示。

这样，注册功能就实现了。但是有个安全问题，账号密码应该被加密存储在数据表中，此时可以通过 bcrypt.js 来解决。在写入数据表的密码中，我们通过 hashSync 进行一层加密：

```
user.username = username;
user.password = bcryptjs.hashSync(password, 10);
const res = await this.userModal.save(user);
```

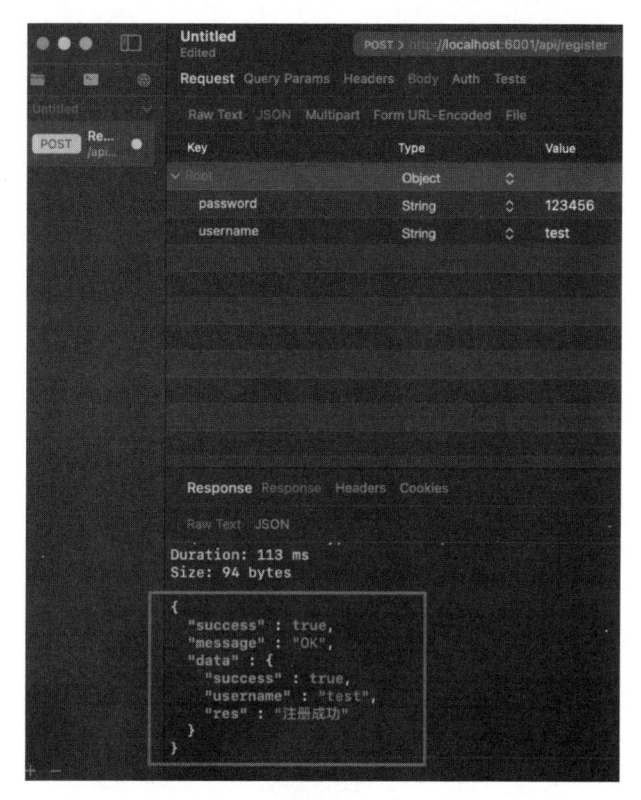

图 6.27

图 6.28

至此，注册接口就开发好了。登录接口的开发与注册接口相差不大，我们实现登录接口的主要逻辑如下：

① 获取请求带来的 username、password；

② 到 user 表中查询用户名，不存在的话就返回异常信息；

③ 通过 bcrypt.js 将登录的明文密码和注册落库的加密密码比较，如果密码错误，则返回异常信息；

第 6 章　全栈化与 Serverless　257

④ 密码正确，登录成功。

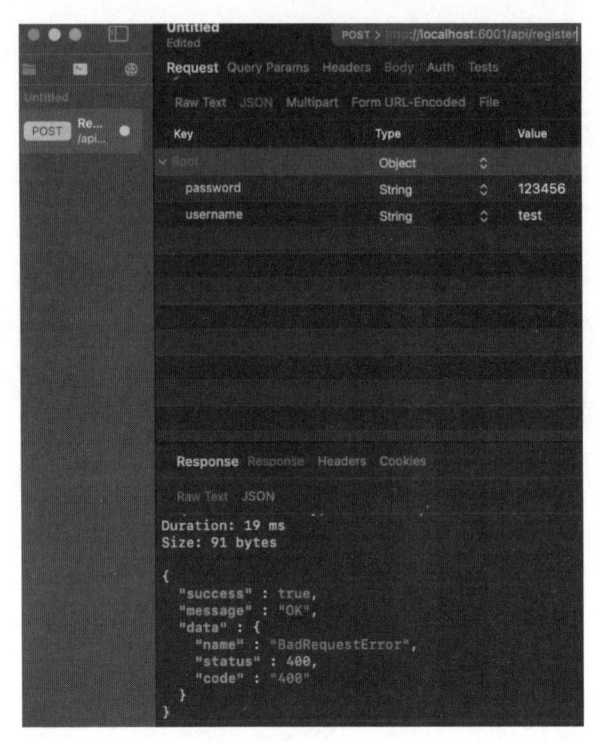

图 6.29

login 接口的代码如下：

```
async login(options: { username: string; password: string }) {
  const { username, password } = options;
  const findRes = await this.userModal.findOne({
    where: {
      username,
    },
  });
    if (!findRes) return new httpError.BadRequestError('不存在该用户');
      const compareRes: boolean = bcryptjs.compareSync(
        password,
        findRes.password
      );
        if (!compareRes)
        return new httpError.BadRequestError('密码错误');
      return {
        success: true
```

```
    };
}
```

接下来我们加入 JWT 鉴权。通过之前对 Nest.js 的 JWT 鉴权流程的讲解，相信读者已经对这个概念和流程非常熟练了，我们用 Midway 来实现一下。

与 Nest 一样，Midway 官方提供了内置 JWT 库，我们安装一下：

```
npm i @midwayjs/jwt --save
```

然后在 configuration.ts 中引入 JWT 组件：

```
import { Configuration, IMidwayContainer } from '@midwayjs/core';
import { IMidwayContainer } from '@midwayjs/core';
import * as jwt from '@midwayjs/jwt';

@Configuration({
  imports: [
    // ...
    jwt,
  ],
})
export class MainConfiguration {
// ...
}
```

接着在 config 中加入 JWT 加密配置信息：

```
// src/config/config.default.ts
export default {
  // ...
  jwt: {
    secret: 'xxxxxxxxxxxxx',
    expiresIn: '2d', // https://github.com/vercel/ms
  },
};
```

配置结束，接下来基于已经实现的注册、登录功能进行改造，主要分两步进行：

- 对于登录接口，验证通过后，产出 token，返回给前端；
- 对于业务接口，依赖 token，做中间件拦截判断鉴权。

先实现第一步，我们只需要在之前的 login 接口中增加 token 的逻辑即可，代码如下：

```
async login(options: { username: string; password: string }) {
```

```
    const { username, password } = options;
    const findRes = await this.userModal.findOne({
      where: {
        username,
      },
    });
      if (!findRes)
      return new httpError.BadRequestError('不存在该用户');
    const compareRes: boolean = bcryptjs.compareSync(
      password,
      findRes.password
    );
    if (!compareRes)
    return new httpError.BadRequestError('密码错误');
    const token = this.jwtService.signSync({ username });
    return {
      accessToken: token,
    };
}
```

当登录成功时，基于用户信息生成加密 token，并返回给前端。前端将其保存在请求头的 authorization 中，每次请求业务接口时都需要将其带给后端，才能返回响应数据。

我们封装一个 jwt.middleware.ts 鉴权中间件，除了登录注册以外，依赖个人账号的业务接口，都先运行到中间件下，代码如下：

```
import { Inject, Middleware, httpError } from '@midwayjs/core';
import { Context, NextFunction } from '@midwayjs/koa';
import { JwtService } from '@midwayjs/jwt';

@Middleware()
export class JwtMiddleware {
  @Inject()
  jwtService: JwtService;

  resolve() {
    return async (ctx: Context, next: NextFunction) => {
      //判断有没有校验信息
      if (!ctx.headers['authorization']) {
        throw new httpError.UnauthorizedError();
      }
      //从 Header 上获取校验信息
```

```
      const parts = ctx.get('authorization').trim().split(' ');
      if (parts.length !== 2) {
        throw new httpError.UnauthorizedError();
      }

      const [scheme, token] = parts;

      if (/^Bearer$/i.test(scheme)) {
        //jwt.verify 方法验证 token 是否有效
        await this.jwtService.verify(token, {
          complete: true,
        });
        await next();
      }
    };
  }

  // 配置忽略认证校验的路由地址
  public match(ctx: Context): boolean {
    const ignore = ['/api/login'].includes(ctx.path);
    return !ignore;
  }
}
```

然后在 configuration.ts 中引入该中间件即可生效：

```
export class ContainerLifeCycle {
  @App()
  app: koa.Application;

  async onReady() {
    // add middleware
    this.app.useMiddleware([JwtMiddleware]);
  }
}
```

除了中间件内部白名单中的接口以外,其他接口都会先运行到 JWT 中间件中。简单测试一下,写一个/getShop 业务接口(不在 JWT 白名单中),首先前端不带 token 发一次请求, 如图 6.30 所示。

结果符合预期：无法访问,被中间件拦截下来了。然后我们调用一次/login 接口,在前端保存一次 token,再带着 token 去请求/getShop 接口,结果如图 6.31 所示。

可以看到,有了 token,业务接口也正常返回报文了。至此,我们便实现了基于 Midway 的注册、登录、鉴权功能。

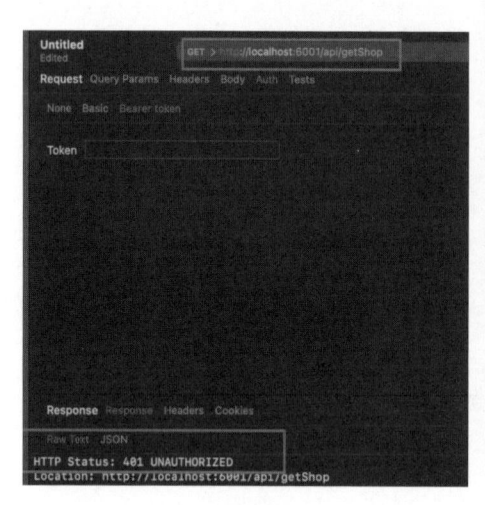

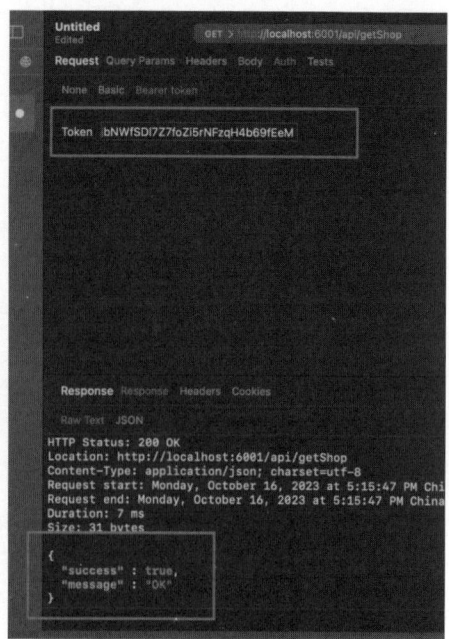

图 6.30

图 6.31

思考题

1. 请描述一下 Nest.js 中 Module、Controller 和 Service 各个模块的作用和关系？

2. 如何在 Nest.js 项目中连接数据库？

3. 什么是 RBAC（role-based access control）模型？用 Nest.js 应该怎么实现？

4. 伪造请求的原理是什么？JWT 是怎么生成和验证的？

5. 如何实现跨端登录？其核心思路是什么？

6. Midway.js 与 Nest.js 的区别是什么？Serverless 有什么优势？

第 7 章
企业级 React 项目实战

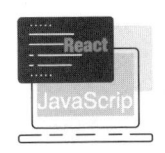

本章主要围绕两个企业级实践展开，分别是：
- React 组件库：介绍如何从 0 到 1 搭建个人的 React 组件库并将其发布到 npm 平台；
- UI 自动化测试平台：基于笔者工作的业务部门，从大厂实际工作需求和思路来讲解如何构建一个覆盖多项目定时测试的平台。

7.1 搭建 React 组件库

要搭建一个 React 组件库，我们需要做两件事：
- 准备一个 Web 项目，也就是官方文档项目，以便在网页中呈现关于组件库的所有教程、案例等；
- 准备一个组件库项目（它可能是一个 src 目录，其中包含了 40 个组件），全部导出，然后将它发布到 npm 中就可以让其他人来使用了。

如何高效地完成这些事呢？本节采用文档站点框架——Dumi.js。它基于 Umi.js 构建，是一款为组件开发场景而生的文档工具，主要用于简化组件库、Hook 库或者其他 React 相关库的文档生成和维护工作。以下是 Dumi 的特点和优势：
- 组件驱动的文档：Dumi 专注于为 React 组件提供开箱即用的文档工具支持，开发者可以直接在 markdown 文件中编写 React 组件，并且利用 Dumi 提供的丰富特性来展示组件的使用示例；
- 基于 Umi 的架构：可以无缝集成 Umi 的路由、构建和插件系统，从而大幅简化文档站点的配置和扩展；
- 主题可定制：Dumi 支持主题定制，允许开发者设计和使用自定义的文档主题，以适应不同的品牌和设计要求；
- 国际化：Dumi 提供了国际化支持，可以帮助开发者构建支持多语言的文档站点；

- TypeScript 支持：Dumi 对 TypeScript 有良好的支持，它可以从 TypeScript 类型中自动生成 API 文档，从而简化文档的维护工作。

接下来就开始组件库的搭建。通过执行 npx create-dumi 即可进入脚手架生成项目模式，选择 React Library 即可生成一个 Dumi 项目，如图 7.1 所示。

执行 yarn start 后，访问 localhost:8080 页面，便实现了一个开箱即用、简洁漂亮的文档项目，并且其头部、logo、右侧 actions、标题都可以进行自定义配置。如图 7.2 所示。

接下来，如何将 React 组件引入页面中呢？可以将每一个组件理解为图 7.3 所示的结构。

Badge 在项目中既是一个组件也是一个路由，而该路由会默认加载目录下的 index.md，我们只需要准备好几个 demo，在 demo 中引用组件（Badge），再统一引入 index.md 中即可。index.md 的例子就像这样：

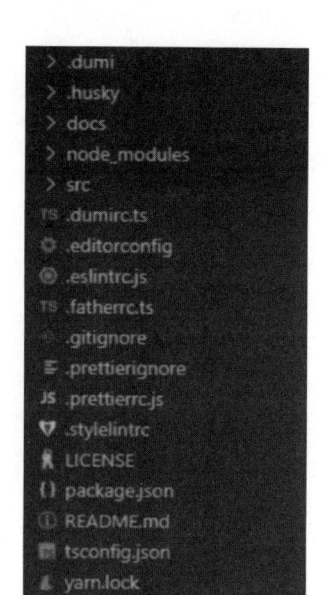

图 7.1

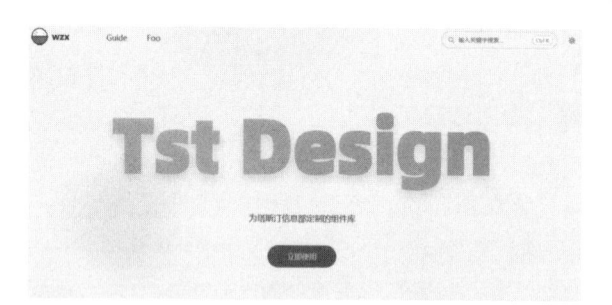

图 7.2

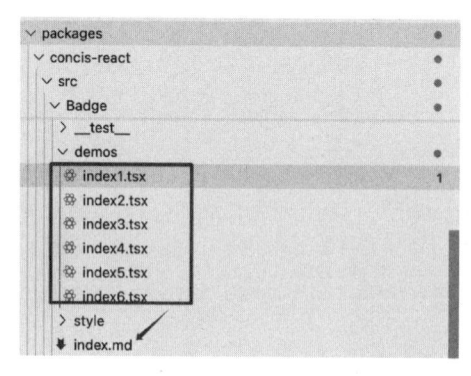

图 7.3

```
---
title: Badge 徽标
nav:
  title: 组件
  path: /common
group:
  title: 数据展示
mobile: false
toc: content
---
```

Badge 徽标
一般出现在图标或文字的右上角。提供及时、重要的信息提示。

基本使用
基础的用法。只需指定 count，即可显示徽标。
<code src="./demos/index1.tsx"></code>

独立使用
children 为空时，将会独立展示徽标。
<code src="./demos/index2.tsx"></code>

小红点
设置 dot，即可只显示小红点而不显示数字。count > 0 时才显示，并通过 offset 自定义小红点位置。
<code src="./demos/index3.tsx"></code>

文本内容
设置 text，可设置徽标为文本内容。
<code src="./demos/index4.tsx"></code>

最大值
设置 maxCount，可以限制最大显示的徽标数值，超过将会加 + 后缀。maxCount 默认为 99。
<code src="./demos/index5.tsx"></code>

API
Name	Description	Type	Default
className	自定义类名	`string`	`--`
style	自定义样式	`CSSProperties`	`--`
count	徽标数字值	`number / ReactNode`	`--`
maxCount	徽标最大值	`number`	`99`

dotStyle	小圆点样式	`CSSProperties`	`{}`
dot	使用小圆点	`boolean`	`false`
offset	小圆点位置值	`Array<number>`	`[2,2]`
text	小圆点文本值	`string`	`''`

在这个例子中，最顶部我们定义了该组件示例文档的标题、路由以及其他信息。在正文中，我们结合 markdown 的能力和 code 组件加载 React 代码块。在最后，我们提供了组件的 props，以表格的形式展示。

整个组件库就是由很多个这样的"套件"组成的。如在源代码目录中可以有这样的 40 个套件来对应 40 个组件。

在发布到 npm 上之前，还需要把源代码全部打包。对此，相比 Webpack, Rollup 是更好的选择方案，有以下几点原因：

- 专注于 ES 模块：Rollup 专门为 ES 模块打包设计，能够生成更高效、更简洁的代码，它通过 tree-shaking 机制去除无用代码，对于减少包体积起到了很大的作用；
- 输出格式：Rollup 能便捷地输出多种格式，包括 ES 模块、CommonJs 模块和 UMD 模块，这使得构建的库可以适用于各种使用场景和模块加载系统；
- 简洁的代码输出：Rollup 默认生成更少的冗余代码和包装代码（boilerplate）。这使得最终的组件库打包文件更小，且不损失运行时（runtime）性能；
- 速度优势：对于库的打包来说，Rollup 能更快地完成任务，尤其是当库中有多个小文件的时候；
- 简单的配置：Rollup 的配置通常比 Webpack 更简单些，对于新手或者偏好"零配置"的开发者而言更加友好。

接下来我们在 Dumi 项目中安装 Rollup，执行命令：

```
npm i rollup -save
```

然后在 src 目录下新建 rollup 构建配置文件，写入一份基础配置代码：

```
// rollup.config.js
export default {
  input: 'src/index.ts',
  output: { file: 'cjs.js', format: 'cjs'}
};
```

在终端执行 rollup -c 即可完成打包。默认在 dist 目录下就可以看到组件库的 commonjs 版本的产物了。为了更灵活地打包组件库，我们可以配置 rollup 插件来支持多样化的功能。比较实用的插件有：

- rollup-plugin-node-resolve：帮助 rollup 查找外部模块，然后导入；
- rollup-plugin-commonjs：将 commonjs 模块转换为 ES2015 供 rollup 处理；
- rollup-plugin-babel：让我们可以使用 ES6 新特性来编写代码；
- rollup-plugin-terser：压缩 JavaScript 代码，包括 ES6 代码；
- rollup-plugin-eslint：js 代码检测。

打包一个组件库项目，使用以上插件已经完全够用了。基于这些插件，我们还可以改造一下配置文件：

```javascript
import resolve from 'rollup-plugin-node-resolve';
import commonjs from 'rollup-plugin-commonjs';
import babel from "rollup-plugin-babel";
import { terser } from 'rollup-plugin-terser';
import { eslint } from 'rollup-plugin-eslint';
export default [
  {
    input: 'src/main.js',
    output: {
      name: 'timeout',
      file: '/lib/tool.js',
      format: 'umd'
    },
    plugins: [
      resolve(), //这样 Rollup 能找到 "ms"
      commonjs(), //这样 Rollup 能转换 "ms" 为一个 ES 模块
      eslint(),
      babel(),
      terser()
    ]
  }
];
```

在 rollup 中，plugin 插件变成了函数式的调用，这样看起来更加清晰。

通常组件库代码需要导出多种模块，对于 commonjs、esm、umd 三种模块的打包，配置也非常简单，只需要配置 output 目录即可：

```javascript
{
  input: 'src/main.js',
  external: ['ms'],
  output: [
    { file: pkg.main, format: 'cjs' },
    { file: pkg.module, format: 'es' },
```

```
      { file: pkg.module, format: 'umd' }
    ]
}
```

file 字段代表了在 package.json 中的对应 key 值，main 字段对应了 commonjs 模块的路径。

组件库打包完成后，就可以直接发布到 npm 上了。开发者需要在 package.json 中配置 files 字段，将项目中的指定目录上传到 npm。如果只上传 dist 目录，那么代码就是这样的：

```
{
  // ...
  "devDependencies": {
    // ...
  },
  "scripts": {
    "build": "NODE_ENV=production rollup -c",
    "dev": "rollup -c -w",
    "test": "node test/test.js",
    "pretest": "npm run build"
  },
  "files": [
    "dist"
  ]
}
```

其次，package.json 中的 version 字段对应了组件库的当前版本。我们需要遵循 semver 语义化版本规范，也就是"主版本号.次版本号.修订版本号"如（1.0.0）：有破坏性的变更，更新第一位；有新功能迭代，更新第二位；bug 修复，变更第三位。这也是大部分开源项目遵循的版本号迭代规范。

发布完的组件库如何在实际项目中使用呢？用户可以直接在其 React 项目中执行类似命令：

```
npm i my-react-library
```

安装以后在页面中导入组件使用即可，就像这样：

```
import { Button } from "my-react-library";
```

无论是搭建官方文档、编写组件，还是将组件发布到 npm 中，以及在实际项目中使用，每个阶段开发者都可以自由扩展。祝读者可以早日拥有属于自己的组件库。

7.2 搭建 UI 自动化测试平台

当部门的业务域非常广，有多条业务线，可能涉及几十甚至上百个项目的时候，通常就需要 UI 自动化测试平台这样的前端基建来保证业务的强交付和高质量。

常规的 UI 自动化测试平台架构一般可分为三层，如图 7.4 所示：

- 业务层：在前端可以添加多个项目，提供测试、生产的域名、项目负责人，对已有项目具备主动执行项目所有测试用例的能力。
- 应用层：存储在自动化测试平台服务端项目中，保存各个项目的测试用例，通过 child_process 来执行 pkg 中的 jest 测试命令。例如要执行 A 项目的测试用例，通过 npm run testA 实现，对应 jest ./autoTest/a。
- 服务层：自动化测试服务端，是所有后台执行脚本能力的聚集点，其核心就是前端调用接口，后端执行测试用例，最后产出测试数据并落库。

图 7.4

笔者推荐的整体技术架构如下：

- 前端：React + Umi + Antd；
- 后端：Node + Midway + Typeorm + Tddl + Jest-puppeteer。

对于技术实现，首先我们需要新建一个后端项目，这里采用 Midway，在终端执行：

```
npm init midway
```

接下来，先搭建测试环境，安装以下依赖包：

```
npm i jest-puppeteer @types/jest jest ts-jest --save -dev
```

下一步就是设计测试用例在项目中的结构，我们在项目根目录中新建一个 testCase 文件夹，再往下划分出一个个项目，里面存放所有的测试用例文件，如图 7.5 所示。

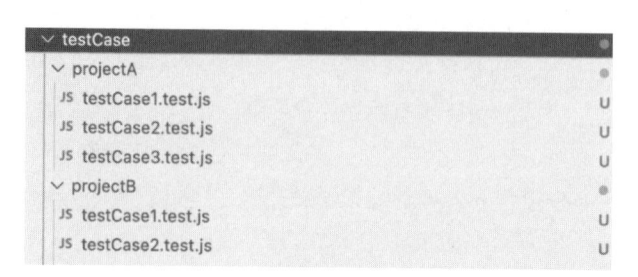

图 7.5

当需要执行指定项目的脚本时，只需要执行对应目录下的所有用例即可，但每个项目都需要一个执行主文件，这里定为 index.test.js，基本代码块如下：

```javascript
require('expect-puppeteer');
const testCase1 = require('./testCase1.test');
const testCase2 = require('./testCase2.test');
const testCase3 = require('./testCase3.test');
const TestConfig = require('./test.config.json');
const projectName = 'projectA';

describe('开始执行创建任务操作', () => {
  beforeAll(async () => {
    await page.goto('http://www.projectA.com');
  });
  it('执行用例1', async () => {
    await testCase1(page);
  });
  it('执行用例2', async () => {
    await testCase2(page);
  });
  it('执行用例3', async () => {
    await testCase3(page);
  });
  afterAll(async () => {
    console.log('用例都执行完啦');
  });
});
```

执行 jest ./testCase/projectA/index.test.js --coverage 就会自动测试 projectA 中的所有测试用例，并产出结果。如此，初步的测试链路就实现了。

自动化测试平台的最终目标是记录测试中的所有数据，所以接下来还需要接入异常感知与拦截等能力。以下是部分常需要记录的数据：

- JS 错误、页面错误、接口请求异常、sourcemap 源代码位置；
- 发生异常时的测试快照；
- 无头浏览器内部操作的完整视频录制等。

在记录这些数据时，由于测试是无状态的，因此需要设计一个监听器，来记录完成一次测试的整个生命周期中遇到的所有需要记录的数据。我们设计一个 RecordService 服务，在测试开始和结束的同时进行该服务的执行和结束，在执行时开启监听，在结束时上报测试记录，代码如下：

```
class RecordService {
  constructor(objcet_id) {
    this.objcet_id = objcet_id; //项目 ID
    this.test_page_count = 0; //测试页面数量
    this.ErrorReducer = new ErrorReducer(); // 记录错误次数
    this.recorder = null;
  }

  async config({ record_id, formInstId }) {
    this.record_id = record_id;
    this.formInstId = formInstId;
  }

  async createPageLoadInfo(param) {
    const url = ''
    const data = {
      objcet_id: this.objcet_id,
      record_id: this.record_id,
      layoutDuration: param.LayoutDuration,
      recalcStyleDuration: param.RecalcStyleDuration,
      scriptDuration: param.ScriptDuration,
      taskDuration: param.TaskDuration,
      whiteDuration: param.WhiteDuration,
      errRequest: param.errRequest,
      jsError: param.jsError,
      url: param.url,
    };
    await makeHttpRequest(url, { method: 'POST', data, dataType: 'text' });
```

```js
  }

  async listenNetwork(page) {
    console.log('listenNetwork start');
    const getRequestInfo = async responseInfo => {
      try {
        const resJson = await responseInfo.json();
        const resData = {
          url: await responseInfo.url(),
          method: await responseInfo.request().method(),
          failure: (await responseInfo.request().failure()?.errorText) || null,
          postData: await responseInfo.request().postData(),
          response: resJson,
          headers: await responseInfo.request().headers(),
          cookies: await page.cookies(),
          success: await resJson?.success,
        };
        return resData;
      } catch (e) {
        console.log('执行请求返回异常', e);
      }
    };

    await page.on('response', async response => {
      if (response.url().indexOf('/h5api') > -1) {
        const res = await getRequestInfo(response);
        if (res?.success) {
          this.ErrorReducer.pushRequest(res);
        } else if (res) {
          this.ErrorReducer.pushErrRequest(res);
        }
      }
    });

    const logStackTrace = async error => {
      this.ErrorReducer.pushJsError(error);
    };

    // 页面崩溃时触发
    page.on('error', logStackTrace);
    // 当页面中的脚本有未捕获异常时发出
```

```
      page.on('pageerror', logStackTrace);
    }

    async screenshot({ imgName, projectName }, page) {
      const path =
        `${process.cwd()}/screenshot/${projectName}/${imgName}.png`;
      await page.screenshot({
        path,
        fullPage: true,
      });
      this.ErrorReducer.pushImgList(path);
    }

    async finish(projectName) {
      const data = {
        objcet_id: this.objcet_id, //项目 ID
        record_id: this.record_id, //记录 ID
        jsError: this.ErrorReducer.jsError,
        errRequest: this.ErrorReducer.errRequest,
        imgList: this.ErrorReducer.imgList.join(','),
      };
    }
}
```

RecordService 服务中共包含了以下五种方法:

● config: 配置测试记录 ID、表单 ID, 用于插入数据表中, 数据由执行脚本时 npm 命令代入;

● createPageLoadInfo: 创建测试执行记录;

● listenNetwork: 核心 API, 包含了 JS 错误、接口异常、页面错误、sourcemap 还原信息的记录, 用于监听测试的整个生命周期;

● screenshot: 异常快照截图;

● finish: 脚本执行结束, 组装所有数据上报接口。

在 RecordService 中引用的 ErrorReducer 类, 在创建的实例中用于保存所有测试数据, 代码如下:

```
class ErrorReducer {
  constructor() {
    this.jsError = [];
    this.errRequest = [];
    this.request = [];
    this.imgList = [];
```

```
  }
  async pushRequest(params) {
    this.request.push(params);
  }
  async pushJsError(params) {
    this.jsError.push(params);
  }
  async pushErrRequest(params) {
    this.errRequest.push(params);
  }
  async pushImgList(params) {
    this.imgList.push(params);
  }
}
```

有了监听器后,我们再回归到之前的测试用例,在测试执行前后进行改造,代码如下:

```
describe('开始执行创建任务操作', () => {
  let exeObj;
  beforeAll(async () => {
    exeObj = global.__AUTOPROJECT__[process.env.npm_config_project_id] =
      new RecordService(process.env.npm_config_project_id);
    exeObj.config({
      record_id: process.env.npm_config_record_id,
      formInstId: process.env.npm_config_forminstid,
    });
    await exeObj.listenNetwork(page);
    await page.goto('http://www.projectA.com');
  });
  // 测试用例...
    afterAll(async () => {
      await exeObj.finish(projectName);
      console.log('用例都执行完啦');
    });
});
```

经过改造后,我们在 finish 方法中已经可以获取到监听器的数据了,并且在执行前传入了项目 ID、测试记录 ID、表单 ID,此时在 finish 中调用新增执行记录接口即可初步实现测试链路。

视频如何录制呢?这里使用了 puppeteer-screen-recorder 库,我们安装一下:

```
npm i puppeteer-screen-recorder --save
```

然后在 RecordService 中加入一个 recordVideo 方法，代码如下：

```
async recordVideo(page, projectName) {
  const screenRecorderOptions = {
    followNewTab: true,
    fps: 25,
    ffmpeg_Path: null,
    videoFrame: {
        width: 1024,
        height: 768,
    },
    videoCrf: 18,
    videoCodec: 'libx264',
    videoPreset: 'ultrafast',
    videoBitrate: 1000,
    autopad: {
      color: 'black' | '#1890ff',
    },
    aspectRatio: '4:3',
  };
  this.recorder = new PuppeteerScreenRecorder(page,
  screenRecorderOptions);
  await this.recorder
    .start(`./video/${projectName}/result.mp4`)
    .then(res => {
      console.log('视频录制开启成功了');
    })
    .catch(err => {
      console.log('视频录制开启失败了', err);
    });
}
```

在测试用例中的 beforeAll 中加入视频录制功能，脚本执行结束后，在项目根目录 video/${projectName} 中即可看到录制完毕的视频。

接下来我们再看一下 sourcemap 该如何还原。因为目前的现状其实是有问题的，通过 puppeteer 拦截到的错误是项目打包后的错误，无法找到报错的代码信息（文件路径、行数、列数），所以我们需要进行映射，这样可以更高效地排查和解决问题。

安装一下实现 sourcemap 还原所需要的依赖包：

```
npm i error-stack-parser source-map-js --save
```

我们先看一下 sourcemap 还原线上代码映射到源码的逻辑, 如图 7.6 所示。

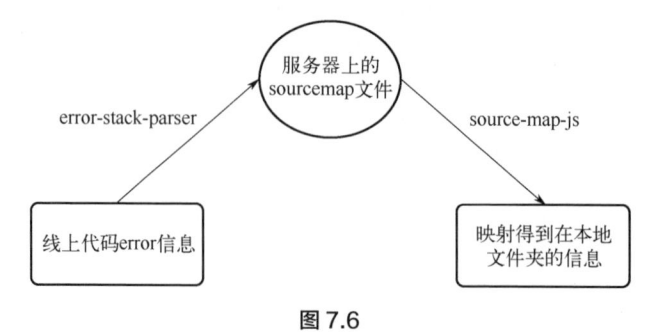

图 7.6

- error-stack-parser 可以基于 js error 类还原出错误信息的堆栈、行数和构建后的报错文件名;
- source-map-js 可以基于线上异常信息和服务器上的 sourcemap 文件来得到最后的源文件信息。

最后我们改装监听器中的 logStackTrace 方法:

```
const logStackTrace = async error => {
  let errorInfo = `错误信息：${error}`;
  // sourcemap 代码映射，获取源代码位置信息
  try {
    const res = ErrorStackParser.parse(new Error(error));
    // 文件名路径分组
    const errorFileNameGroup = res[0].fileName.split('/');
    // 线上版本, 0.0.160
    const version = errorFileNameGroup.find(_ => _.includes('0.0')) || '';
    // 文件名 xxx.js
    const fileName = errorFileNameGroup[errorFileNameGroup.length - 1];
    const aoneNameIndex = errorFileNameGroup.findIndex(_ =>
      _.includes('eleme')
    );
    // 项目名, projectA
    const projectName = errorFileNameGroup[aoneNameIndex + 1];
    if (version && fileName && aoneName && projectName) {
        const sourceMapPath =
    `https://sourcemap.def.alibaba-inc.com/ sourcemap/${aoneName}/${projectName}/${version}/client/js/${fileName}.map`;
    let sourceRes = await loadSourceMap(sourceMapPath);
    if (sourceRes.includes('Redirecting to')) {
```

```
      // sourcemap 的文件在 OSS，需要重定向请求一次
      let ossPath = sourceRes.split('Redirecting to')[1].trim();
      ossPath = ossPath.slice(0, ossPath.length - 1);
      sourceRes = await loadSourceMap(ossPath);
    }
    const sourceData = JSON.parse(sourceRes);
    const consumer = await new sourceMap.SourceMapConsumer(sourceData);
    const result = consumer.originalPositionFor({
      line: Number(res[0].lineNumber),
      column: Number(res[0].columnNumber),
    });
    errorInfo += `，文件路径：${result.source}，报错代码行数：${result.line}
行，报错代码列数：${result.column}列`;
  }
} catch (e) {
  console.log('捕捉 sourcemap 出错：', e);
} finally {
  this.ErrorReducer.pushJsError(errorInfo);
}
};
```

代码块中主要是企业获取 sourcemap 的思路逻辑。如果是普通项目，对于 sourcemap 文件没有保护机制的话，直接通过网络请求访问就可以了。

目前我们已实现了测试用例、运行脚本、测试时数据整合、异常监听，最后还需要一个平台来运行脚本，手动执行脚本接口。此时把测试命令抽离到接口里，暴露出去即可。

首先定义一个 runScript 接口：

```
@Provide()
export class ScriptService {
  async runScript(options: ProjectOptions) {
    const resData = await runScript(options);
    return resData;
  }
}
```

然后实现 runScript 的逻辑，主要思路就是接收项目 ID，执行指定项目的 jest 终端命令，最后把监听器的数据上报到数据库里。如果有异常，可选择通过钉钉服务等告警（结合具体需求而定）。具体代码如下：

```
const runScript = async (options: ProjectOptions) => {
  const { pid } = options;
```

```
    const yidaService = await useInject(YidaService);
    const dingdingService = await useInject(DingDingService);
    const uploadService = await useInject(UploadService);
    const { pid } = options;

    // 根据pid查scriptName，代码略过
    return new Promise(async resolve => {
      if (scriptName) {
        //新增记录，获取 record_id
        console.log('project_id', pid);
        //新建一条执行记录初始数据
        await exec(
          `tnpm run ${scriptName} --project_id=${pid}
          --record_id=${record_id} --formInstId=${result}
          --host=${hostPath}`, // 项目参数代入
          async (_err, _stdout, _stderr) => {
            await yidaService.updateProject({
              formInstId,
              textField_l7fxypdj: '', //解除当前项目正在执行状态
              textareaField_l7gcpy7t: _stderr,
            });
            const { passed, total, time } = splitTestResult(_stderr);
            //将完整数据更新到当次执行记录中，代码略过
            //如果有异常，调用钉钉服务告警，代码略过
            resolve({
              _err,
              _stdout,
              _stderr,
            });
          }
        );
      } else {
        resolve({});
      }
    });
}
```

最后我们建立一个前端项目，技术栈使用 React 即可，可考虑制作一个项目首页，类似于图 7.7 所示。

至此，UI 自动化测试平台的主要逻辑思路就讲完了。限于篇幅，本节对于其他功能及很多具体细节未能展开，比如数据库、数据表逻辑，图片视频上传 OSS，项目、执行记录的接口等内容。对此，建议读者结合自身或企业需求，在实践中

去积极探索。

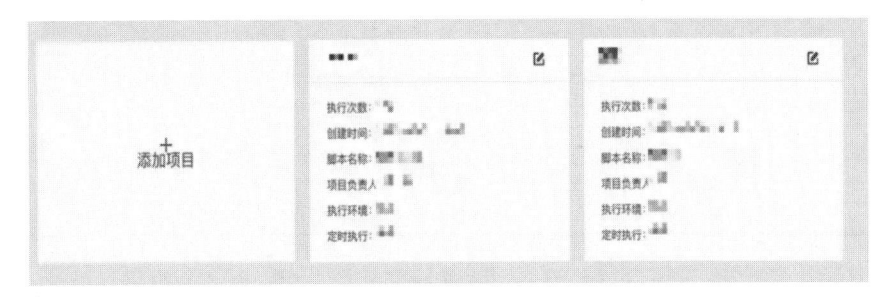

图 7.7

思考题

1. React 组件库是什么？市面上有哪些比较热门的相似的库？
2. 如何搭建一个 React 组件库？在项目架构中，对于组件库的 core 部分和文档站点部分，应该如何分离、独立部署？
3. 如何将组件库发布到 npm 中并下载使用？
4. 组件库的版本号迭代管理应当遵循怎样的原则？
5. UI 自动化测试平台的核心实现思路是什么？
6. 如果想在 UI 自动化测试平台中直接在线编写测试用例，同时保留项目中测试用例文件的原子能力，应该怎样设计迭代？

结语

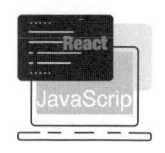

随着前端开发领域的不断发展，React 作为一种流行的 JavaScript 库将继续发挥重要作用。未来的挑战可能包括提高性能、优化用户体验以及处理日益复杂的应用程序架构。此外，Serverless 技术和微服务架构也将成为关注焦点，因为它们可以帮助开发者更加专注于业务逻辑而不是基础设施管理。

为了进一步提升自己的技能，感兴趣的读者可以考虑以下几方面进行发展性学习：

（1）深入研究 React 生态系统

除了官方文档外，还有许多优秀的博客文章、视频教程以及开源项目可供参考。例如，可以探索 React Native 以了解如何使用 React 构建原生移动应用程序。

（2）实践项目经验

参与实际项目是提高技能的最佳途径之一，尝试寻找机会参与到开源项目或个人项目中，这样可以在实践中不断成长。

（3）阅读优秀源码

例如 ahooks 的源码。读者可以寻找更多的主流 React 生态库了解其内部工作机制，这有助于加深对于 React 的理解，花时间去阅读和分析优秀源码是成长的不错选择。

（4）持续关注新技术

前端开发领域发展迅速，也许 React 也会更新到更高版本，而新工具和框架仍将层出不穷，保持对最新趋势的关注，并适时更新自己的知识体系非常重要。

总之，在掌握本书内容后，请继续深化技能、始终关注行业动态，祝愿你成为一名优秀的前端工程师。